Yearbook of Astronomy 2018

Front Cover: Image of Comet C/2014 Q2 (Lovejoy) captured by Damian Peach on 20 January 2015. See the article *Comets and How to Photograph Them* for further details.

YEARBOOK OF
ASTRONOMY
2018

Brian Jones
Richard S Pearson FRAS

WHITE
OWL

First published in Great Britain in 2017
by Pen & Sword White Owl
An imprint of Pen & Sword Books Limited
47 Church Street
Barnsley
South Yorkshire
S70 2AS

ISBN 978 1 52671 741 2

Typeset in Ehrhardt
by Mac Style Ltd, Bridlington, East Yorkshire

Printed and bound in India
by Replika Press Pvt. Ltd.

Pen & Sword Books Limited incorporates the imprints of Atlas,
Archaeology, Aviation, Discovery, Family History, Fiction, History,
Maritime, Military, Military Classics, Politics, Select, Transport,
True Crime, Air World, Frontline Publishing, Leo Cooper,
Remember When, Seaforth Publishing, The Praetorian Press, Wharncliffe
Local History, Wharncliffe Transport,
Wharncliffe True Crime and White Owl

For a complete list of Pen & Sword titles please contact
PEN & SWORD BOOKS LIMITED
47 Church Street, Barnsley, South Yorkshire, S70 2AS, England
E-mail: enquiries@pen-and-sword.co.uk
Website: www.pen-and-sword.co.uk

Contents

Article Section

Miscellaneous

Editors' Foreword

The Yearbook of Astronomy has long been an indispensable publication both for the active observer of the skies and the armchair astronomer, and the annual appearance of this excellent book has been eagerly anticipated by astronomers, both amateur and professional, for well over half a century. As a number of you will be aware, the future of the Yearbook was under threat following the decision to make the 2016 edition the last. However, the series was rescued, both through the publication of a special 2017 edition (which successfully maintained the continuity of the Yearbook) and a successful search for a publisher to take this iconic publication on and help us take it to even greater heights as the Yearbook of Astronomy approaches its diamond jubilee in 2022.

Although a limited edition, copies of the Yearbook of Astronomy 2017 are still available to purchase via **www.starlight-nights.co.uk/subscriber-2017-yearbook-astronomy**

The content of the current edition is described below. Over time new topics and themes will be introduced into the Yearbook to allow it to keep pace with the increasing range of skills, techniques and observing methods now open to amateur astronomers, this in addition to articles relating to our rapidly-expanding knowledge of the Universe in which we live. There will be an interesting mix, some articles written at a level which will appeal to the casual reader and some at what may be loosely described as a more academic level. To sum up, the intention is to fully maintain (and increase) the usefulness and relevance of the Yearbook of Astronomy to the interests of the readership who are, without doubt, the most important aspect of the Yearbook and the reason it exists in the first place.

The first section of the 2018 edition is devoted to the Monthly Star Charts which have been drawn up by David Harper and which show the night sky throughout the year. Two sets of twelve charts are provided, one set for observers in the Northern Hemisphere and one set for those in the Southern

Hemisphere. Each pair of charts depict the entire sky as two semi-circular half-sky views, one looking north and the other looking south. These are followed by the *Monthly Sky Notes* which have been compiled by Lynne Stockman and which provide details of the positions and visibility of the planets throughout 2018. Each section of the Sky Notes is accompanied by a short article, the range of which includes items on space exploration and astronomy as well as observing guides to specific constellations on view at the time.

The first part of the book is rounded off by lists of *Phases of the Moon in 2018* and *Eclipses in 2018* together with a trio of articles by Neil Norman, these being *Comets in 2018*, *Minor Planets in 2018* and *Meteor Showers in 2018*, all three titles being fairly self-explanatory describing as they do the occurrence and visibility of examples of these three classes of object during and throughout the year.

Summaries of *Solar System Exploration in 2017* and *Astronomy in 2017* have been compiled by Peter Rea and Rod Hine. The former covers a range of topics including the demise of the two-decade long Cassini mission at Saturn following the probe being intentionally crashed into the planet, as well as an exciting and tantalizing look to the future and the retargeting of the New Horizons spacecraft towards an encounter with the Kuiper Belt Object 2014 MU69 on 1 January 2019. With its intended target being located over 1.6 billion kilometres (1 billion miles) beyond the orbit of Pluto, the New Horizons spacecraft is taking our exploration of the Solar System to what may literally be termed as 'new horizons'.

In his article *Astronomy in 2017* Rod Hine covers a good range of topics, including the latest in the search for exoplanets, of which over 3,600 had been located at time of writing. The first confirmed detection of an extrasolar planet came almost 20 years ago and the number of known exoplanets continues to grow. Taking this a stage further, and bearing in mind that planets within our own Solar System play host to a total of over 170 moons, it is by no means unreasonable to think that there are moons orbiting other planets. Space probes have shown us that some of the natural satellites orbiting planets in our own backyard have the potential to harbour life, and this makes the detection of other moons, beyond the confines of our Solar System, a highly-sought-after goal for astronomers.

Amongst the topics covered by Neil Haggath in his article *Anniversaries in 2018* is the 350th anniversary of Isaac Newton making the first reflecting telescope in 1668. Although the principle of reflecting telescopes had been described a few years previously, Newton was the first to actually make one. It is interesting to note that Newton's first reflector had an aperture of just 1.5 inches when we consider that 2018 also marks the 150th anniversary of the birth of the American astronomer George Ellery Hale in 1868. Although Hale's achievements were many, he is best remembered as a founder of observatories and a builder of great telescopes.

Such has been the impact of astrophotography on amateur astronomy over recent years that we intend to feature much more on this subject in future Yearbooks. This edition contains the article *Comets and How to Photograph Them* by Damian Peach whose exploits in imaging the night sky are of a very high quality. The depiction of Comet C/2014 Q2 (Lovejoy) featured on the cover of the current volume was provided by Damian and illustrates his abilities in astrophotography very well indeed.

John McCue introduces us to some of the wonders of double stars in his article *Double and Multiple Stars*. After drawing the distinctions between the different types of double star he tells us something of the history of double star observation as well as drawing on personal experiences to convey his obvious fascination for this subject.

Moving back to the historical side of astronomy, *Some Pioneering Lady Astronomers* by Mike Frost throws some light on the achievements of several notable women astronomers including the much-travelled historian, astronomy writer and accomplished solar observer Mary Acworth Evershed, née Orr. Her marriage to fellow solar observer John Evershed in 1906 was followed by John being offered the post of director of the Kodaikanal observatory in India from where Mary was able to pursue her observational interests. She later became the first director of the BAA Historical Section following its formation in 1930.

Is There Still a Place for Art in Astronomy? is the question posed by David Hardy, who introduces us to the subject by drawing on the works of past luminaries in the field such as Scriven Bolton, Lucien Rudaux and the 'old master' Chesley Bonestell. We have all seen examples of space art used to good effect in the past and, as David points out, artists have now been presented with what is perhaps their most exciting area of work yet following the seemingly-relentless discovery of extrasolar worlds or 'exoplanets'. Today's artists continue to show us how space and its exploration will be developed, and they can depict the surfaces and environments of alien worlds in ways that are impossible to achieve by more conventional means, such as space probes. Space art can be said to have come of age and it seems clear that its future is assured.

David Harland introduces us to *Supermassive Black Holes* and explains the discovery of supermassive black holes in the cores of galaxies, including our own, and the manner in which they evolved from initially voraciously swallowing up material to ultimately starving and entering a prolonged period of dormancy. This is followed by our final article *Modern Video Astronomy* in which Steve Wainwright informs us how video astronomy involves the use of video techniques to produce live images of astronomical objects, even deep sky objects, on monitor screens for real-time viewing. Steve looks into the history of video astronomy as well as bringing us up-to-date with its application by modern day astronomers and observers.

The final section of the book starts off with *Some Interesting Variable Stars* by Roger Pickard which contains useful information on variables, including the well known long-period variable star Mira, as well as predictions for timings of minimum brightness for the famous eclipsing binary Algol for 2018. *Some Interesting Double Stars* and *Some Interesting Nebulae, Clusters and Galaxies* present a selection of objects for you to seek out in the night sky. The lists are by no means definitive and may well omit your favourite celestial targets. If this is the case, please let us know and we will endeavour to add to the lists in future editions of the Yearbook.

The book rounds off with a selection of *Astronomical Organizations*, which lists organizations and associations across the world through which you can further pursue your interest and participation in astronomy (if there are any that we have omitted please let us know) and *Our Contributors*, which contains brief background details of the numerous writers who have contributed to the Yearbook of Astronomy 2018. Finally we have introduced what we believe to be a first for the Yearbook of Astronomy, this being a *Glossary*, in which brief but informative explanations are given for many of the words and much of the terminology used in the current edition (as well as a few more).

Grateful thanks go to those individuals who have contributed their time and effort to the Yearbook of Astronomy 2018, included in this number being Garry (Garfield) Blackmore who has produced the artwork for the smaller star charts included with a number of the articles. Our grateful thanks also go to Jonathan Wright, Lori Jones, Lauren Burton and Janet Brookes of Pen &

Sword Books Ltd. Without their help, belief and confidence in the Yearbook of Astronomy, the 2017 edition of this iconic publication may well have been the last.

Brian Jones - Editor
Bradford, West Riding of Yorkshire

Richard S. Pearson - Editor
Nottingham, Nottinghamshire

May 2017

Preface

The information given in this edition of the Yearbook of Astronomy is in narrative form. The positions of the planets given in the Monthly Sky Notes often refer to the constellations in which they lie at the time. These can be found on the star charts which collectively show the whole sky via two charts depicting the northern and southern circumpolar stars and forty-eight charts depicting the main stars and constellations for each month of the year. The northern and southern circumpolar charts show the stars that are within 45° of the two celestial poles, while the monthly charts depict the stars and constellations that are visible throughout the year from Europe and North America or from Australia and New Zealand. The monthly charts overlap the circumpolar charts. Wherever you are on the Earth, you will be able to locate and identify the stars depicted on the appropriate areas of the chart(s).

There are numerous star atlases available that offer more detailed information, such as Sky & Telescope's POCKET SKY ATLAS and Norton's STAR ATLAS and Reference Handbook to name but a couple. In addition, more precise information relating to planetary positions and so on can be found in a number of publications, a good example of which is The Handbook of the British Astronomical Association, as well as many of the popular astronomy magazines such as the British monthly periodicals Sky at Night and Astronomy Now and the American monthly magazines Astronomy and Sky & Telescope.

About Time

Before the late 18th century, the biggest problem affecting mariners sailing the seas was finding their position. Latitude was easily determined by observing the altitude of the pole star above the northern horizon. Longitude, however, was far more difficult to measure. The inability of mariners to determine their longitude often led to them getting lost, and on many occasions shipwrecked. To address this problem King Charles II established the Royal Observatory at Greenwich in 1675 and from here, Astronomers Royal began the process of measuring and cataloguing the stars as they passed due south across the Greenwich meridian.

Now mariners only needed an accurate timepiece (the chronometer invented by Yorkshire-born clockmaker John Harrison) to display GMT (Greenwich Mean Time). Working out the local standard time onboard ship and subtracting this from GMT gave the ship's longitude (west or east) from the Greenwich meridian. Therefore mariners always knew where they were at sea and the longitude problem was solved.

Astronomers use a time scale called Universal Time (UT). This is equivalent to Greenwich Mean Time and is defined by the rotation of the Earth. The Yearbook of Astronomy gives all times in UT rather than in the local time for a particular city or country. Times are expressed using the 24-hour clock, with the day beginning at midnight, denoted by 00:00. Universal Time (UT) is related to local mean time by the formula:

Local Mean Time = UT - west longitude

In practice, small differences in longitude are ignored and the observer will use local clock time which will be the appropriate Standard (or Zone) Time. As the formula indicates, places in west longitude will have a Standard Time slow on UT, while those in east longitude will have a Standard Time fast on UT. As examples we have:

Standard Time in

New Zealand	UT +12 hours
Victoria, NSW	UT +10 hours
Western Australia	UT + 8 hours
South Africa	UT + 2 hours
British Isles	UT
Eastern Standard Time	UT – 5 hours
Central Standard Time	UT – 6 hours
Pacific Standard Time	UT – 8 hours

During the periods when Summer Time (also called Daylight Saving Time) is in use, one hour must be added to Standard Time to obtain the appropriate Summer/Daylight Saving Time. For example, Pacific Daylight Time is UT – 7 hours.

Using the Yearbook of Astronomy as an Observing Guide

Notes on the Monthly Star Charts

The star charts on the following pages show the night sky throughout the year. There are two sets of charts, one for use by observers in the Northern Hemisphere and one for those in the Southern Hemisphere. The first set is drawn for latitude 52°N and can be used by observers in Europe, Canada and most of the United States. The second set is drawn for latitude 35°S and show the stars as seen from Australia and New Zealand. Twelve pairs of charts are provided for each of these latitudes.

Each pair of charts shows the entire sky as two semi-circular half-sky views, one looking north and the other looking south. A given pair of charts can be used at different times of year. For example, chart 1 shows the night sky at midnight on 21 December, but also at 2am on 21 January, 4am on 21 February and so forth. The accompanying table will enable you to select the correct chart for a given month and time of night. The caption next to each chart also lists the dates and times of night for which it is valid.

The charts are intended to help you find the more prominent constellations and other objects of interest mentioned in the monthly observing notes. To avoid the charts becoming too crowded, only stars of magnitude 4.5 or brighter are shown. This corresponds to stars that are bright enough to be seen from any dark suburban garden on a night when the Moon is not too close to full phase.

Each constellation is depicted by joining selected stars with lines to form a pattern. There is no official standard for these patterns, so you may occasionally find different patterns used in other popular astronomy books for some of the constellations.

Any map projection from a sphere onto a flat page will by necessity contain some distortions. This is true of star charts as well as maps of the Earth. The distortion on the half-sky charts is greatest near the semi-circular boundary of each chart, where it may appear to stretch constellation patterns out of shape.

The charts also show selected deep-sky objects such as galaxies, nebulae and star clusters. Many of these objects are too faint to be seen with the naked eye, and you will need binoculars or a telescope to observe them. Please refer to *Some Interesting Nebulae, Star Clusters and Galaxies* (page 277) for more information.

Selecting the Correct Charts

The table opposite shows which of the charts to use for particular dates and times throughout the year and will help you to select the correct pair of half-sky charts for any combination of month and time of night.

The Earth takes 23 hours 56 minutes (and 4 seconds) to rotate once around its axis with respect to the fixed stars. Because this is around four minutes shorter than a full 24 hours, the stars appear to rise and set about 4 minutes earlier on each successive day, or around an hour earlier each fortnight. Therefore, as well as showing the stars at 10pm (22h in 24-hour notation) on 21 January, chart 1 also depicts the sky at 9pm (21h) on 6 February, 8pm (20h) on 21 February and 7pm (19h) on 6 March.

The times listed do not include summer time (daylight saving time), so if summer time is in force you must subtract one hour to obtain standard time (GMT if you are in the United Kingdom) before referring to the chart. For example, to find the correct chart for mid-September in the northern hemisphere at 3am summer time, first of all subtract one hour to obtain 2am (2h) standard time. Then you can consult the table, where you will find that you should use chart 11.

The table does not indicate sunrise, sunset or twilight. In northern temperate latitudes, the sky is still light at 18h and 6h from April to September, and still light at 20h and 4h from May to August. In Australia and New Zealand, the sky is still light at 18h and 6h from October to March, and in twilight (with only bright stars visible) at 20h and 04h from November to January.

Local Time	18h	20h	22h	0h	2h	4h	6h
January	11	12	1	2	3	4	5
February	12	1	2	3	4	5	6
March	1	2	3	4	5	6	7
April	2	3	4	5	6	7	8
May	3	4	5	6	7	8	9
June	4	5	6	7	8	9	10
July	5	6	7	8	9	10	11
August	6	7	8	9	10	11	12
September	7	8	9	10	11	12	1
October	8	9	10	11	12	1	2
November	9	10	11	12	1	2	3
December	10	11	12	1	2	3	4

Legend to the Star Charts

STARS

Symbol	Magnitude
•	0 or brighter
•	1
•	2
•	3
·	4
·	5

| • | Double star |
| ◉ | Variable star |

DEEP-SKY OBJECTS

Symbol	Type of object
✵	Open star cluster
◌	Globular star cluster
□	Nebula
▦	Cluster with nebula
○	Planetary nebula
↘	Galaxy
	Magellanic Clouds

Star Names

There are over 200 stars with proper names, most of which are of Roman, Greek or Arabic origin although only a couple of dozen or so of these names are used regularly. Examples include Arcturus in Boötes, Castor and Pollux in Gemini and Rigel in Orion.

A system whereby Greek letters were assigned to stars was introduced by the German astronomer and celestial cartographer Johann Bayer in his star atlas Uranometria, published in 1603. Bayer's system is applied to the brighter stars within any particular constellation, which are given a letter from the Greek alphabet followed by the genitive case of the constellation in which the star is located. This genitive case is simply the Latin form meaning 'of' the constellation. Examples are the stars Alpha Boötis and Beta Centauri which translate literally as 'Alpha of Boötes' and 'Beta of the Centaur'.

As a general rule, the brightest star in a constellation is labelled Alpha (α), the second brightest Beta (β), and the third brightest Gamma (γ) and so on, although there are some constellations where the system falls down. An example is Gemini where the principal star (Pollux) is designated Beta Geminorum, the second brightest (Castor) being known as Alpha Geminorum.

There are only 24 letters in the Greek alphabet (see below), the consequence of which was that the fainter naked eye stars needed an alternative system of classification. The system in popular use is that devised by the first Astronomer Royal John Flamsteed in which the stars in each constellation are listed numerically in order from west to east. Although many of the brighter stars within any particular constellation will have both Greek letters and Flamsteed numbers, the latter are generally used only when a star does not have a Greek letter.

The Greek Alphabet

α	Alpha	ι	Iota	ρ	Rho
β	Beta	κ	Kappa	σ	Sigma
γ	Gamma	λ	Lambda	τ	Tau
δ	Delta	μ	Mu	υ	Upsilon
ε	Epsilon	ν	Nu	φ	Phi
ζ	Zeta	ξ	Xi	χ	Chi
η	Eta	ο	Omicron	ψ	Psi
θ	Theta	π	Pi	ω	Omega

The Names of the Constellations

On clear, dark, moonless nights, the sky seems to teem with stars although in reality you can never see more than a couple of thousand or so at any one time when looking with the unaided eye. Each and every one of these stars belongs to a particular constellation, although the constellations that we see in the sky, and which grace the pages of star atlases, are nothing more than chance alignments. The stars that make up the constellations are often situated at vastly differing distances from us and only appear close to each other, and form the patterns that we see, because they lie in more or less the same direction as each other as seen from Earth.

A large number of the constellations are named after mythological characters, and were given their names thousands of years ago. However, those star groups lying close to the south celestial pole were discovered by Europeans only during the last few centuries, many of these by explorers and astronomers who mapped the stars during their journeys to lands under southern skies. This resulted in many of the newer constellations having modern-sounding names, such as Octans (the Octant) and Microscopium (the Microscope), both of which were devised by the French astronomer Nicolas Louis De La Caille during the early 1750s.

Over the centuries, many different suggestions for new constellations have been put forward by astronomers who, for one reason or another, felt the need to add new groupings to star charts and to fill gaps between the traditional constellations. Astronomers drew up their own charts of the sky, incorporating their new groups into them. A number of these new constellations had cumbersome names, notable examples including Officina Typographica (the Printing Shop) introduced by the German astronomer Johann Bode in 1801; Sceptrum Brandenburgicum (the Sceptre of Brandenburg) introduced by the German astronomer Gottfried Kirch in 1688; Taurus Poniatovii (Poniatowski's Bull) introduced by the Polish-Lithuanian astronomer Martin Odlanicky Poczobut in 1777; and Quadrans Muralis (the Mural Quadrant) devised by the French astronomer Joseph-Jerôme de Lalande in 1795. Although these have long since been rejected, the latter has been immortalised by the annual Quadrantid meteor shower, the radiant of which lies in an area of sky formerly occupied by Quadrans Muralis.

During the 1920s the International Astronomical Union (IAU) systemised matters by adopting an official list of 88 accepted constellations, each with official spellings and abbreviations. Precise boundaries for each constellation were then drawn up so that every point in the sky belonged to a particular constellation.

The abbreviations devised by the IAU each have three letters which in the majority of cases are the first three letters of the constellation name, such as AND for Andromeda, EQU for Equuleus, HER for Hercules, ORI for Orion and so on. This trend is not strictly adhered to in cases where confusion may arise. This happens with the two constellations Leo (abbreviated LEO) and Leo Minor (abbreviated LMI). Similarly, because Triangulum (TRI) may be mistaken for Triangulum Australe, the latter is abbreviated TRA. Other instances occur with Sagitta (SGE) and Sagittarius (SGR) and with Canis Major (CMA) and Canis Minor (CMI) where the first two letters from the second names of the constellations are used. This is also the case with Corona Australis (CRA) and Corona Borealis (CRB) where the first letter of the second name of each constellation is incorporated. Finally, mention must be made of Crater (CRT) which has been abbreviated in such a way as to avoid confusion with the aforementioned CRA (Corona Australis).

The table shown on the following pages contains the name of each of the 88 constellations together with the translation and abbreviation of the constellation name. The constellations depicted on the monthly star charts are identified with their abbreviations rather than the full constellation names.

The Constellations

Andromeda	Andromeda	AND
Antlia	The Air Pump	ANT
Apus	The Bird of Paradise	APS
Aquarius	The Water Carrier	AQR
Aquila	The Eagle	AQL
Ara	The Altar	ARA
Aries	The Ram	ARI
Auriga	The Charioteer	AUR
Boötes	The Herdsman	BOO
Caelum	The Graving Tool	CAE

Camelopardalis	The Giraffe	CAM
Cancer	The Crab	CNC
Canes Venatici	The Hunting Dogs	CVN
Canis Major	The Great Dog	CMA
Canis Minor	The Little Dog	CMI
Capricornus	The Goat	CAP
Carina	The Keel	CAR
Cassiopeia	Cassiopeia	CAS
Centaurus	The Centaur	CEN
Cepheus	Cepheus	CEP
Cetus	The Whale	CET

Chamaeleon	The Chameleon	CHA	Monoceros	The Unicorn	MON
Circinus	The Pair of Compasses	CIR	Musca	The Fly	MUS
			Norma	The Level	NOR
Columba	The Dove	COL	Octans	The Octant	OCT
Coma Berenices	Berenice's Hair	COM	Ophiuchus	The Serpent Bearer	OPH
Corona Australis	The Southern Crown	CRA	Orion	Orion	ORI
Corona Borealis	The Northern Crown	CRB	Pavo	The Peacock	PAV
			Pegasus	Pegasus	PEG
Corvus	The Crow	CRV	Perseus	Perseus	PER
Crater	The Cup	CRT	Phoenix	The Phoenix	PHE
Crux	The Cross	CRU	Pictor	The Painter's Easel	PIC
Cygnus	The Swan	CYG	Pisces	The Fish	PSC
Delphinus	The Dolphin	DEL	Piscis Austrinus	The Southern Fish	PSA
Dorado	The Goldfish	DOR	Puppis	The Stern	PUP
Draco	The Dragon	DRA	Pyxis	The Mariner's Compass	PYX
Equuleus	The Foal	EQU			
Eridanus	The River	ERI	Reticulum	The Net	RET
Fornax	The Furnace	FOR	Sagitta	The Arrow	SGE
Gemini	The Twins	GEM	Sagittarius	The Archer	SGR
Grus	The Crane	GRU	Scorpius	The Scorpion	SCO
Hercules	Hercules	HER	Sculptor	The Sculptor	SCL
Horologium	The Pendulum Clock	HOR	Scutum	The Shield	SCT
			Serpens Caput and Cauda	The Serpent	SER
Hydra	The Water Snake	HYA			
Hydrus	The Lesser Water Snake	HYI	Sextans	The Sextant	SEX
			Taurus	The Bull	TAU
Indus	The Indian	IND	Telescopium	The Telescope	TEL
Lacerta	The Lizard	LAC	Triangulum	The Triangle	TRI
Leo	The Lion	LEO	Triangulum Australe	The Southern Triangle	TRA
Leo Minor	The Lesser Lion	LMI			
Lepus	The Hare	LEP	Tucana	The Toucan	TUC
Libra	The Scales	LIB	Ursa Major	The Great Bear	UMA
Lupus	The Wolf	LUP	Ursa Minor	The Little Bear	UMI
Lynx	The Lynx	LYN	Vela	The Sail	VEL
Lyra	The Lyre	LYR	Virgo	The Virgin	VIR
Mensa	The Table Mountain	MEN	Volans	The Flying Fish	VOL
Microscopium	The Microscope	MIC	Vulpecula	The Fox	VUL

The Monthly Star Charts

Northern Hemisphere Star Charts

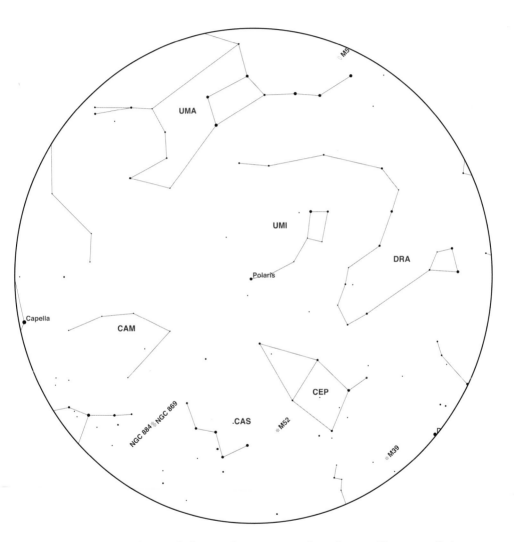

This chart shows stars lying at declinations between +45 and +90 degrees. These constellations are circumpolar for observers in Europe and North America.

1N

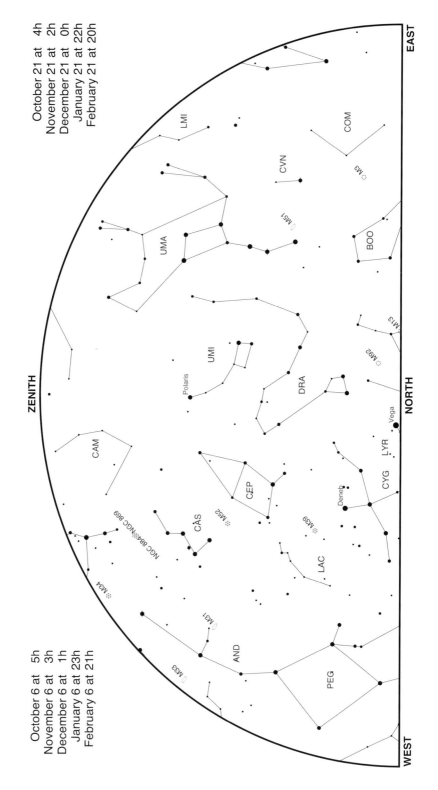

October 21 at 4h
November 21 at 2h
December 21 at 0h
January 21 at 22h
February 21 at 20h

October 6 at 5h
November 6 at 3h
December 6 at 1h
January 6 at 23h
February 6 at 21h

EAST

ZENITH

NORTH

WEST

LMI
COM
CVN
M3
M51
UMA
BOO
M13
M92
UMI
Polaris
DRA
Vega
CAM
CEP
LYR
CAS
M52
CYG
Deneb
NGC 884/NGC 889
M39
LAC
M34
M31
AND
M33
PEG

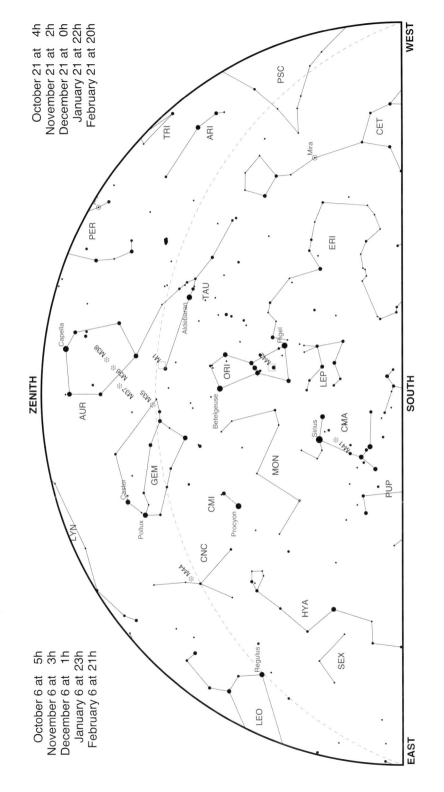

1S

WEST

PSC

CET

Mira

TRI

ARI

PER

ERI

TAU

Aldebaran

Capella

M38

M36

M37

M35

M1

AUR

ORI

Rigel

M42

Betelgeuse

LEP

ZENITH

Castor

GEM

Pollux

MON

CMA

Sirius

M41

PUP

SOUTH

LYN

CMI

Procyon

CNC

M44

HYA

LEO

Regulus

SEX

EAST

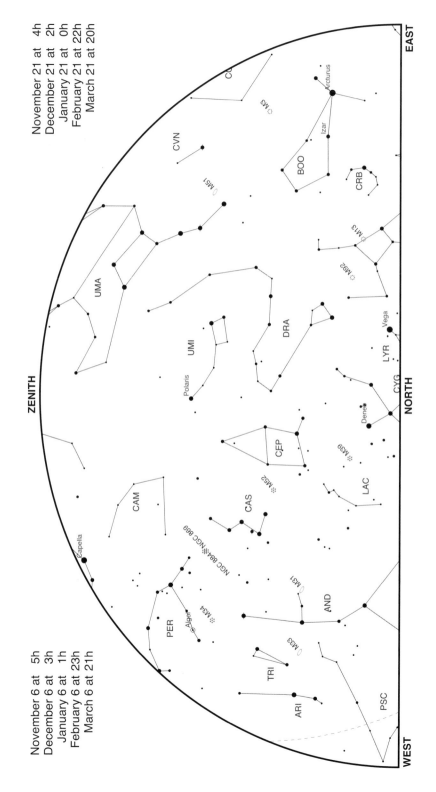

2N

November 6 at 5h
December 6 at 3h
January 6 at 1h
February 6 at 23h
March 6 at 21h

EAST

ZENITH

NORTH

WEST

CO

M3

Arcturus

Izar

BOO

CRB

CVN

M51

M13

M92

UMA

DRA

Vega

UMI

LYR

Polaris

CYG

Deneb

CEP

M39

CAS

M52

LAC

CAM

Capella

NGC 884 NGC 869

PER

M34

Algol

M31

AND

M33

TRI

PSC

ARI

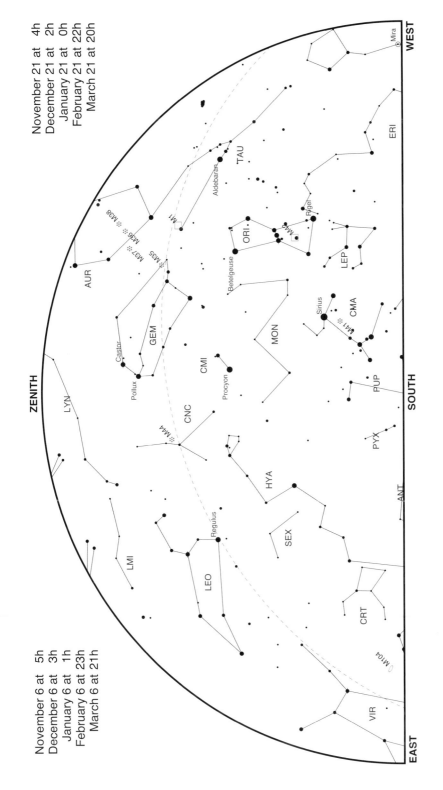

2S

November 6 at 5h
December 6 at 3h
January 6 at 1h
February 6 at 23h
March 6 at 21h

WEST

ZENITH

SOUTH

EAST

Mira

ERI

TAU

Aldebaran

ORI.

Rigel

M42

LEP

Betelgeuse

M1

M35

M36

M37

M38

AUR

Castor

Pollux

GEM

LYN

CMI

Procyon

MON

Sirius

M41

CMA

PUP

PYX

ANT

CNC

M44

LMI

Regulus

LEO

HYA

SEX

CRT

M104

VIR

3N

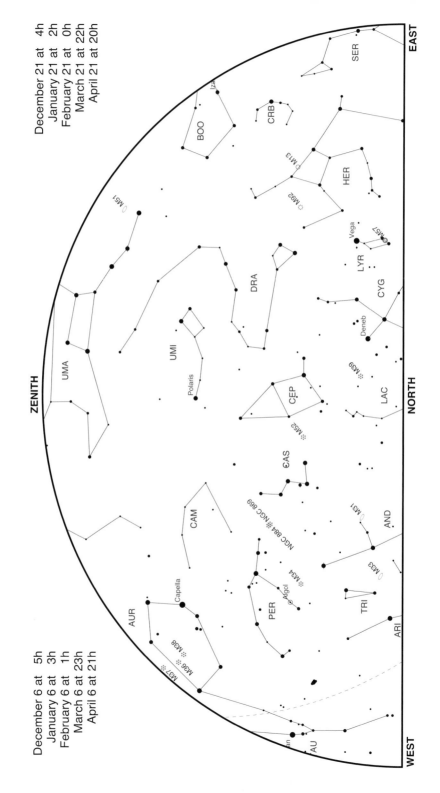

December 6 at 5h
January 6 at 3h
February 6 at 1h
March 6 at 23h
April 6 at 21h

EAST

WEST

NORTH

ZENITH

SER

BOO

CRB

HER

M13

M92

LYR

Vega

M57

CYG

Deneb

M51

DRA

UMI

Polaris

CEP

M52

LAC

M39

UMA

CAS

CAM

NGC 884 NGC 869

M31

AND

M33

M34

PER

Algol

TRI

ARI

AUR

Capella

M38

M36

M37

AU

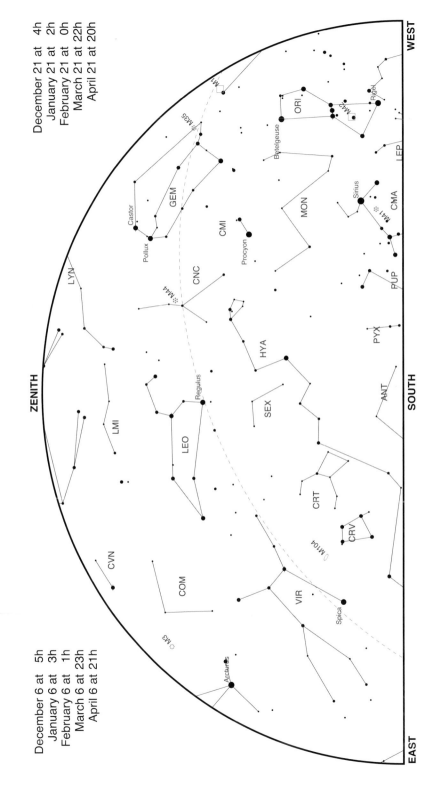

December 21 at 4h
January 21 at 2h
February 21 at 0h
March 21 at 22h
April 21 at 20h

WEST

ZENITH

EAST

SOUTH

December 6 at 5h
January 6 at 3h
February 6 at 1h
March 6 at 23h
April 6 at 21h

Castor
Pollux
GEM
LYN
LMI
CVN
COM
M3
Arcturus
VIR
Spica
M104
CRV
CRT
SEX
HYA
ANT
PYX
PUP
CMA
M41
Sirius
MON
Betelgeuse
ORI
M42
Rigel
LEP
M35
M44
CNC
CMI
Procyon
Regulus
LEO

4N

January 21 at 4h
February 21 at 2h
March 21 at 0h
April 21 at 22h
May 21 at 20h

January 6 at 5h
February 6 at 3h
March 6 at 1h
April 6 at 23h
May 6 at 21h

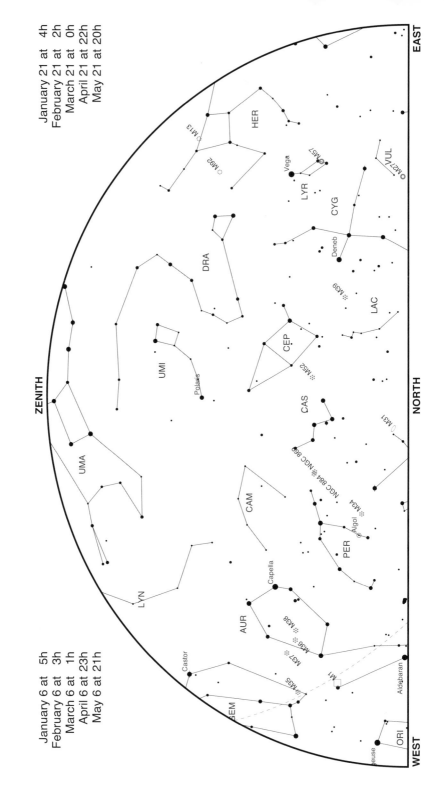

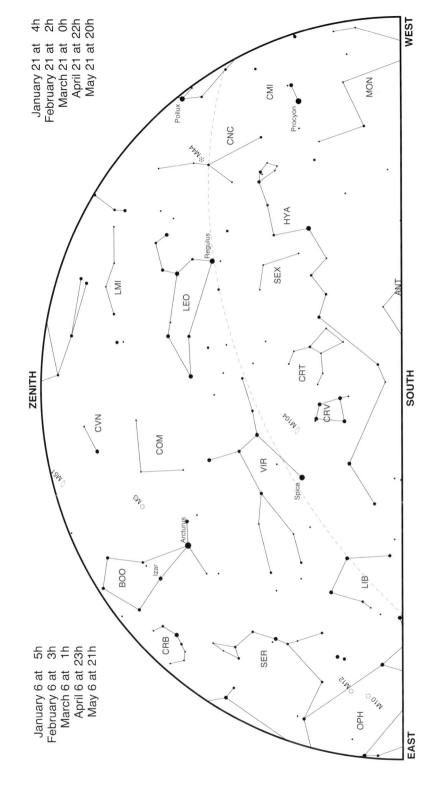

January 21 at 4h
February 21 at 2h
March 21 at 0h
April 21 at 22h
May 21 at 20h

January 6 at 5h
February 6 at 3h
March 6 at 1h
April 6 at 23h
May 6 at 21h

WEST

4S

ZENITH

SOUTH

EAST

Pollux
CMI
MON
Procyon
CNC
M44
HYA
SEX
Regulus
LMI
LEO
ANT
CVN
COM
CRT
M51
M3
CRV
M104
Izar
Arcturus
VIR
Spica
BOO
LIB
CRB
SER
M12
M10
OPH

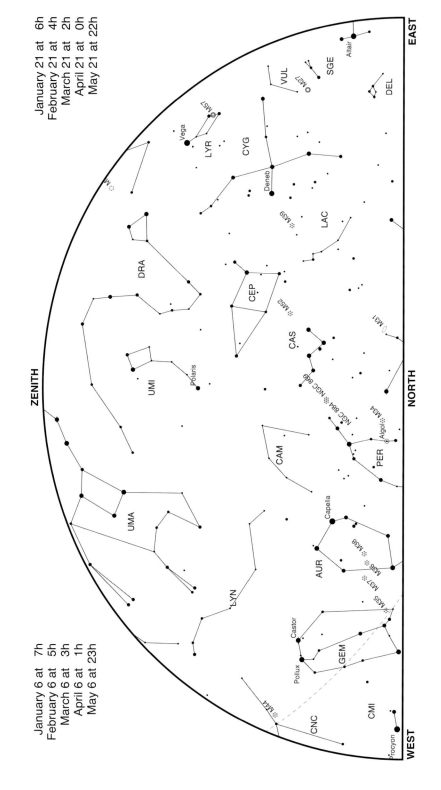

5N

EAST

WEST

NORTH

ZENITH

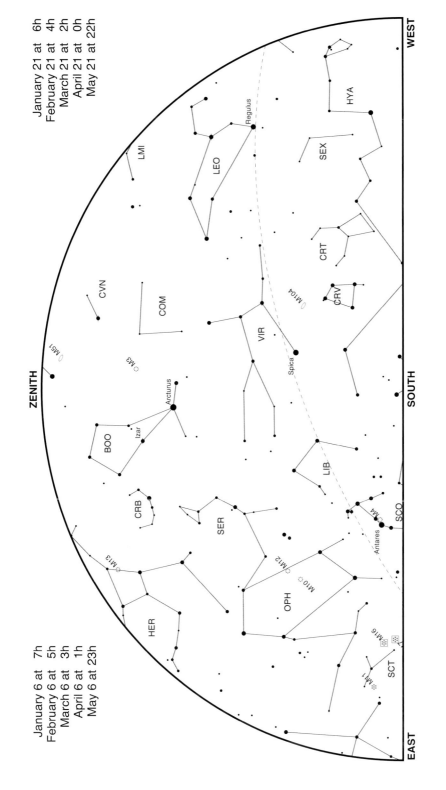

January 21 at 6h
February 21 at 4h
March 21 at 2h
April 21 at 0h
May 21 at 22h

January 6 at 7h
February 6 at 5h
March 6 at 3h
April 6 at 1h
May 6 at 23h

WEST

EAST

SOUTH

ZENITH

LMI

LEO

Regulus

HYA

SEX

CRT

CRV

M104

VIR

Spica

CVN

COM

M3

M51

Arcturus

Izar

BOO

CRB

SER

LIB

M13

HER

M5

OPH

M12

M10

SCO

Antares

M4

M16

M11

SCT

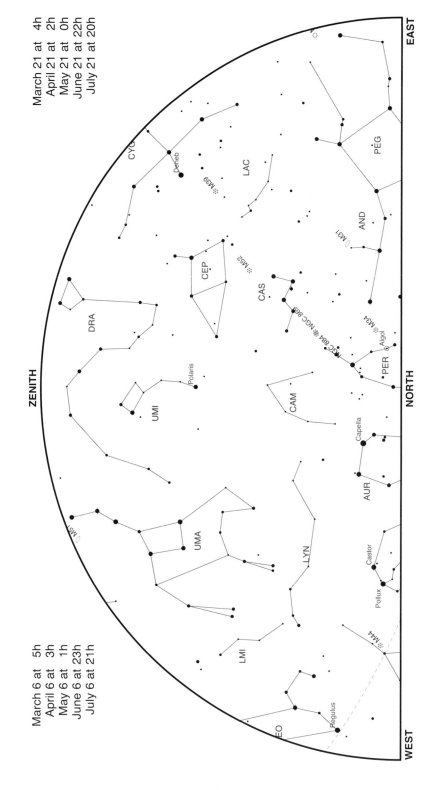

6N

March 6 at 5h
April 6 at 3h
May 6 at 1h
June 6 at 23h
July 6 at 21h

ZENITH

EAST

WEST

NORTH

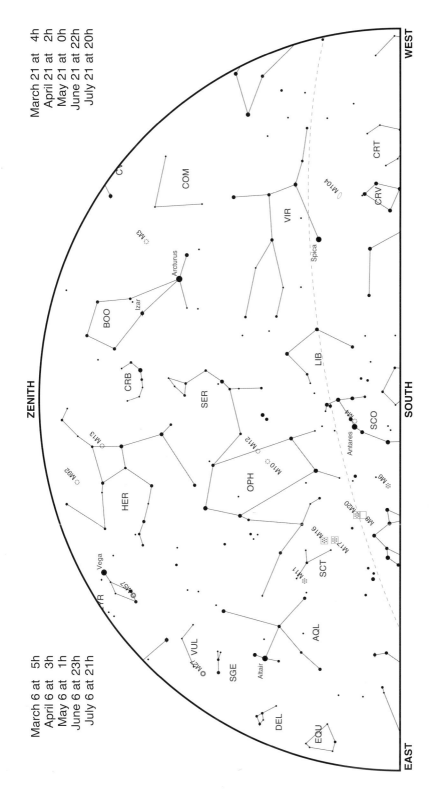

6S

March 21 at 4h
April 21 at 2h
May 21 at 0h
June 21 at 22h
July 21 at 20h

March 6 at 5h
April 6 at 3h
May 6 at 1h
June 6 at 23h
July 6 at 21h

WEST

ZENITH

EAST

SOUTH

CVN

COM

VIR

CRT

CRV

M104

Spica

M3

Arcturus

Izar

BOO

LIB

CRB

SER

M13

M92

HER

OPH

M10

M12

SCO

Antares

M4

M6

M20

M8

M17

M16

SCT

M11

Vega

LYR

M57

VUL

M27

SGE

AQL

Altair

DEL

EQU

7N

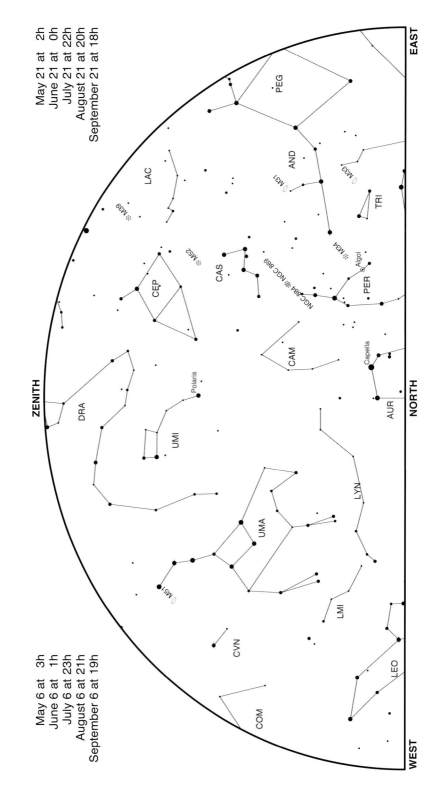

May 21 at 2h
June 21 at 0h
July 21 at 22h
August 21 at 20h
September 21 at 18h

May 6 at 3h
June 6 at 1h
July 6 at 23h
August 6 at 21h
September 6 at 19h

EAST

ZENITH

WEST

NORTH

PEG

LAC

AND

M31

M33

TRI

M34

M39

M52

CAS

Algol

PER

NGC 884 NGC 869

CEP

CAM

Capella

DRA

AUR

Polaris

UMI

LYN

UMA

M51

LMI

CVN

LEO

COM

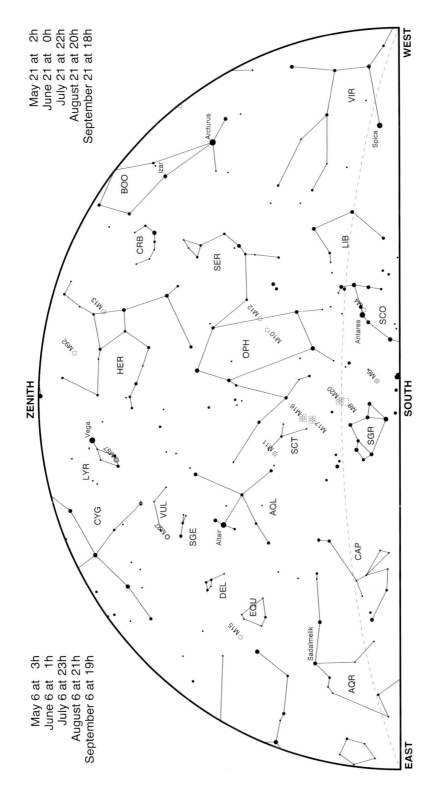

7S

May 21 at 2h
June 21 at 0h
July 21 at 22h
August 21 at 20h
September 21 at 18h

May 6 at 3h
June 6 at 1h
July 6 at 23h
August 6 at 21h
September 6 at 19h

WEST

EAST

SOUTH

ZENITH

VIR
Spica
BOO
Arcturus
Izar
CRB
SER
LIB
M13
M92
HER
OPH
M10
M12
SCO
Antares
M6
Vega
LYR
M57
CYG
VUL
M27
SGE
SCT
M11
M17
M16
SGR
M8
M20
CAP
AQL
Altair
DEL
EQU
M15
AQR
Sadalmelik

8N

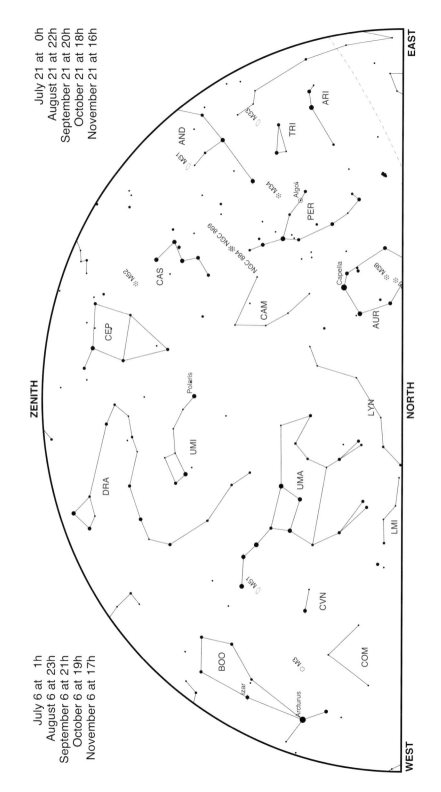

July 21 at 0h
August 21 at 22h
September 21 at 20h
October 21 at 18h
November 21 at 16h

July 6 at 1h
August 6 at 23h
September 6 at 21h
October 6 at 19h
November 6 at 17h

EAST

ZENITH

NORTH

WEST

AND
M31
M33
TRI
ARI
M34
Algol
PER
NGC 884 NGC 869
CAS
M52
CAM
AUR
Capella
M38
M36
CEP
Polaris
UMI
LYN
DRA
UMA
LMI
M51
CVN
COM
BOO
Izar
Arcturus
M3

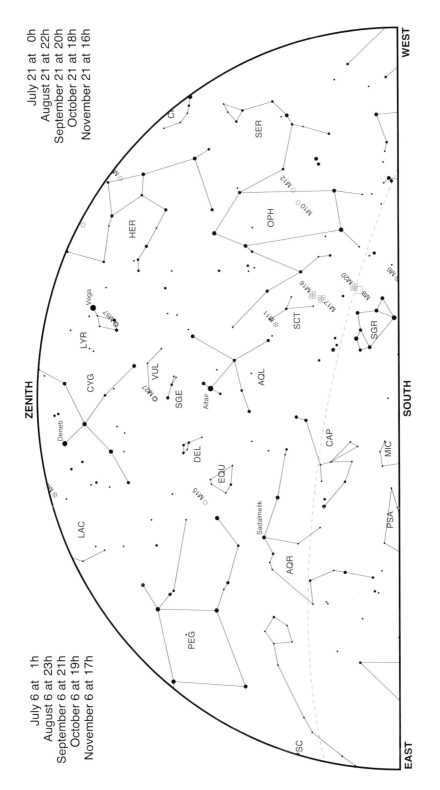

8S

WEST

July 21 at 0h
August 21 at 22h
September 21 at 20h
October 21 at 18h
November 21 at 16h

ZENITH

EAST

July 6 at 1h
August 6 at 23h
September 6 at 21h
October 6 at 19h
November 6 at 17h

SOUTH

SER

OPH

M12

M10

M16

M20

M8

M17

M6

SGR

M11

SCT

HER

Vega

M57

LYR

CYG

VUL

M27

SGE

Altair

AQL

Deneb

DEL

EQU

CAP

MIC

M15

LAC

PEG

Sadalmelik

AQR

PSA

SC

9N

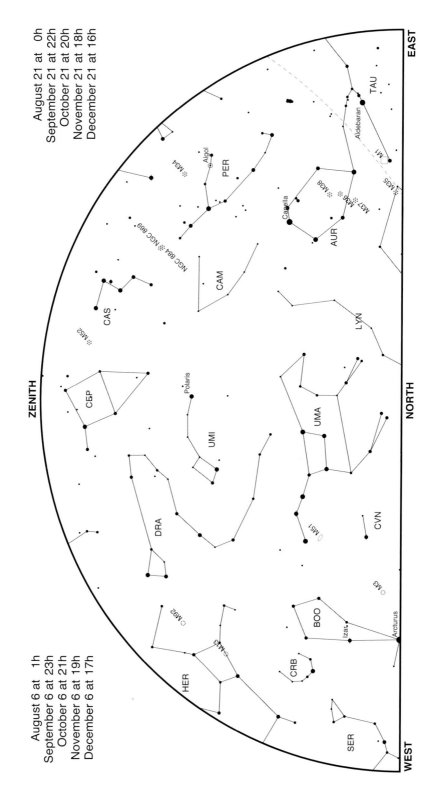

August 6 at 1h
September 6 at 23h
October 6 at 21h
November 6 at 19h
December 6 at 17h

EAST

ZENITH

NORTH

WEST

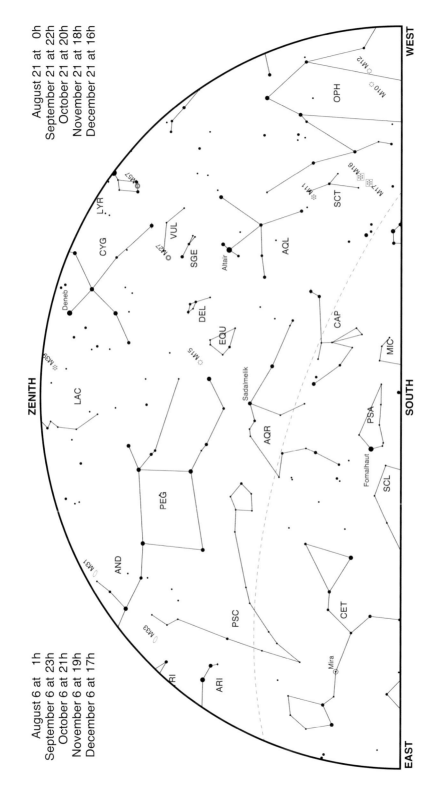

9S

August 21 at 0h
September 21 at 22h
October 21 at 20h
November 21 at 18h
December 21 at 16h

August 6 at 1h
September 6 at 23h
October 6 at 21h
November 6 at 19h
December 6 at 17h

WEST

EAST

SOUTH

ZENITH

OPH
M10
M12
M57
LYR
CYG
VUL
M27
SGE
Altair
AQL
SCT
M11
M17 M16
Deneb
DEL
EQU
CAP
MIC
M15
M39
LAC
Sadalmelik
AQR
PSA
Fomalhaut
SCL
PEG
AND
M31
M33
TRI
ARI
PSC
CET
Mira

10N

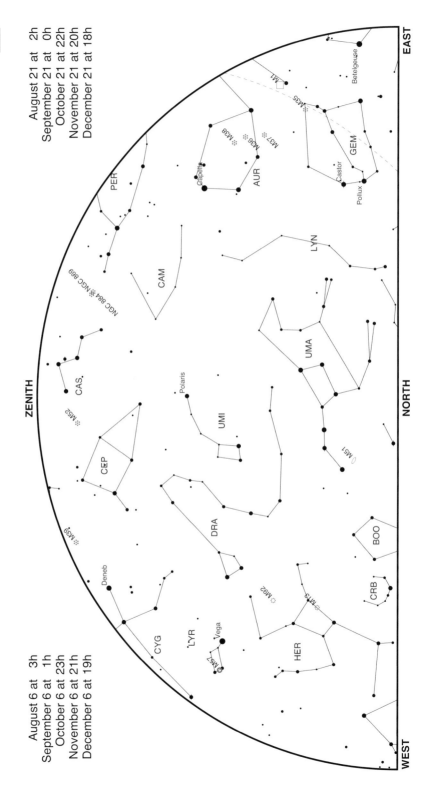

August 21 at 2h
September 21 at 0h
October 21 at 22h
November 21 at 20h
December 21 at 18h

August 6 at 3h
September 6 at 1h
October 6 at 23h
November 6 at 21h
December 6 at 19h

EAST

WEST

NORTH

ZENITH

PER

CAM

CAS.

CEP

CYG

LYR

DRA

UMI

UMA

LYN

AUR

GEM

HER

BOO

CRB

Polaris

Deneb

Vega

Capella

Castor

Pollux

Betelgeuse

M38

M36

M37

M35

M1

M52

NGC 884 NGC 869

M39

M92

M13

M57

M51

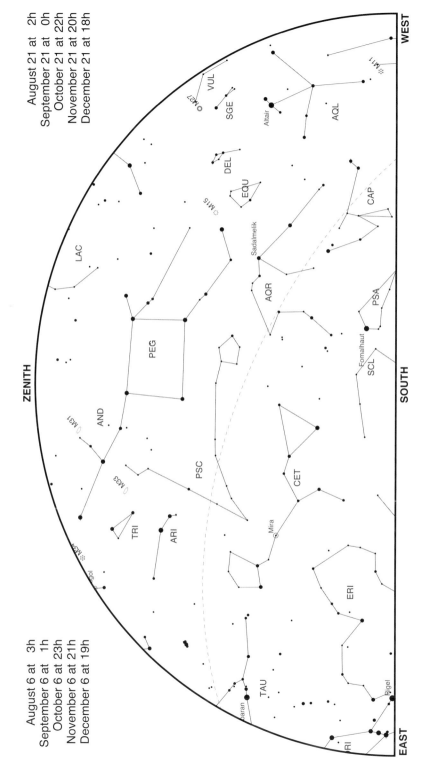

10S

August 6 at 3h
September 6 at 1h
October 6 at 23h
November 6 at 21h
December 6 at 19h

WEST

ZENITH

EAST

SOUTH

VUL
SGE
M27
DEL
EQU
M15
LAC
AQL
Altair
CAP
Sadalmelik
AQR
PSA
Fomalhaut
SCL
PEG
AND
M31
M33
PSC
TRI
ARI
Algol
NGC
CET
Mira
ERI
TAU
Daran
Rigel
RI

11N

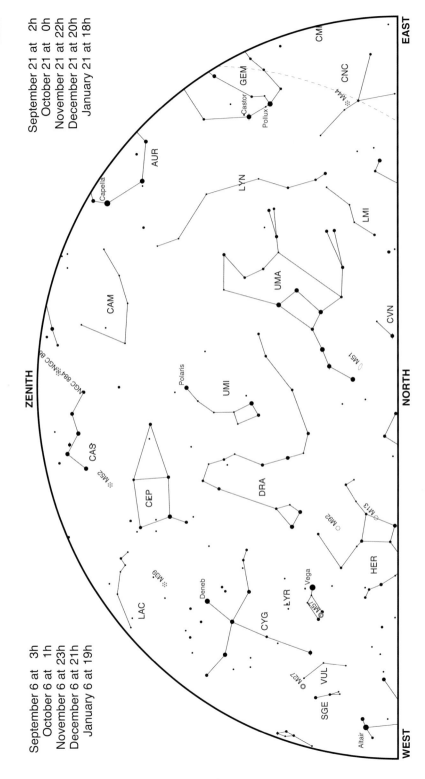

September 21 at 2h
October 21 at 0h
November 21 at 22h
December 21 at 20h
January 21 at 18h

September 6 at 3h
October 6 at 1h
November 6 at 23h
December 6 at 21h
January 6 at 19h

EAST

NORTH

WEST

ZENITH

CMI
GEM
Castor
Pollux
CNC
M44
AUR
Capella
LYN
LMI
CAM
UMA
CVN
M51
NGC 884
NGC 869
Polaris
UMI
DRA
M92
M13
CAS
M52
CEP
HER
LAC
M39
Deneb
CYG
LYR
Vega
M57
VUL
SGE
M27
Altair

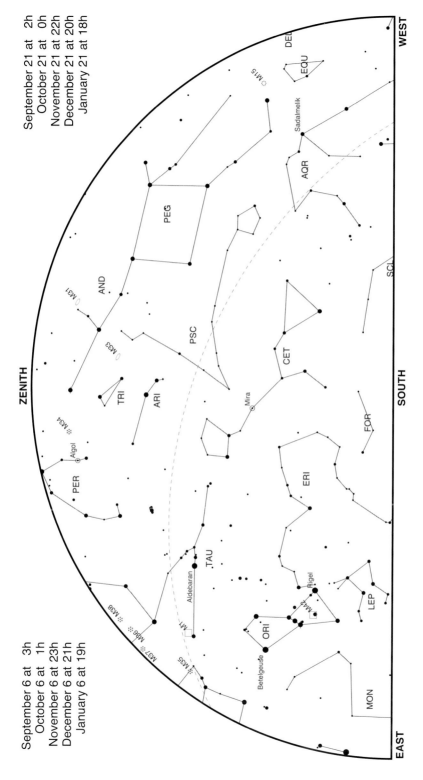

September 21 at 2h
October 21 at 0h
November 21 at 22h
December 21 at 20h
January 21 at 18h

September 6 at 3h
October 6 at 1h
November 6 at 23h
December 6 at 21h
January 6 at 19h

ZENITH

WEST

11S

EAST

SOUTH

12N

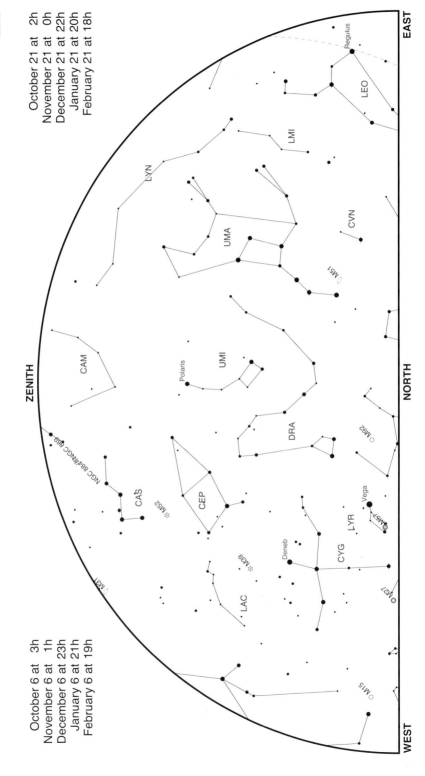

October 21 at 2h
November 21 at 0h
December 21 at 22h
January 21 at 20h
February 21 at 18h

October 6 at 3h
November 6 at 1h
December 6 at 23h
January 6 at 21h
February 6 at 19h

EAST

ZENITH

NORTH

WEST

Regulus

LEO

LMI

LYN

CVN

M51

UMA

CAM

Polaris

UMI

DRA

M92

CAS

M52

CEP

M57

Vega

LYR

Deneb

M39

CYG

LAC

M27

M15

NGC 884+NGC 869

M31

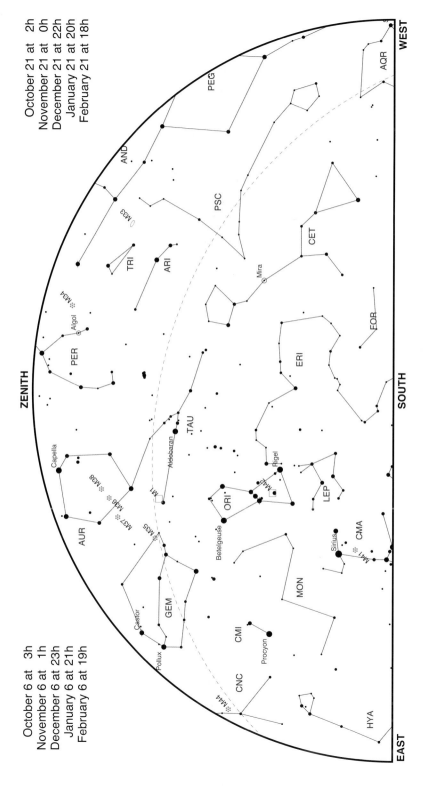

12S

October 21 at 2h
November 21 at 0h
December 21 at 22h
January 21 at 20h
February 21 at 18h

October 6 at 3h
November 6 at 1h
December 6 at 23h
January 6 at 21h
February 6 at 19h

WEST

EAST

ZENITH

SOUTH

AQR

PEG

AND

PSC

CET

TRI

ARI

Mira

M33

Algol

PER

M34

FOR

ERI

Capella

TAU

Aldebaran

M38

AUR

M36

M37

ORI

Rigel

M42

LEP

Betelgeuse

M35

M1

CMA

Sirius

M41

Castor

GEM

Pollux

MON

CMI

Procyon

CNC

M44

HYA

Southern Hemisphere Star Charts

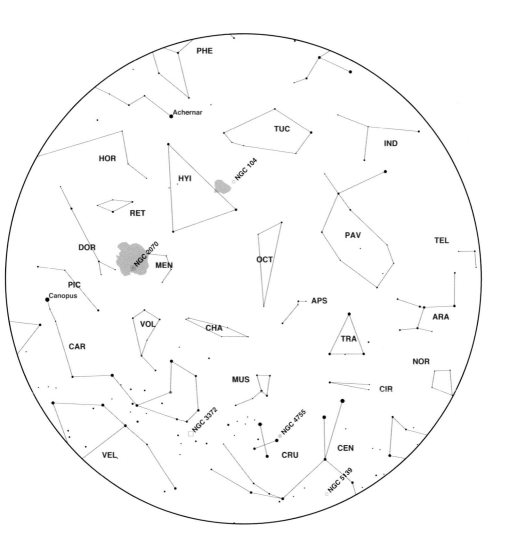

This chart shows stars lying at declinations between –45 and –90 degrees. These constellations are circumpolar for observers in Australia and New Zealand.

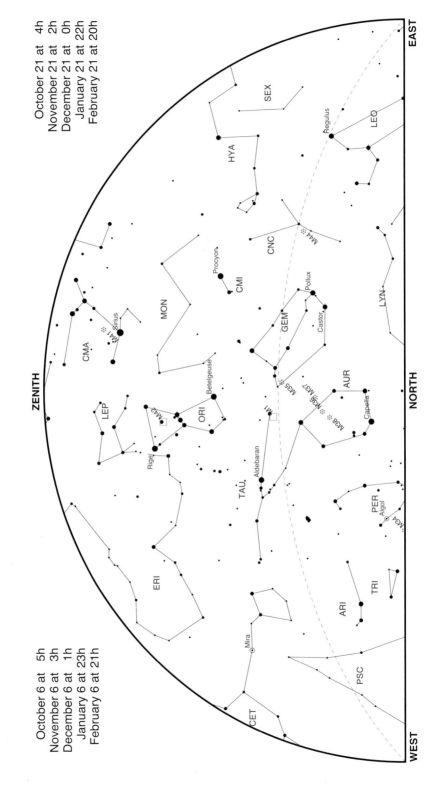

October 21 at 4h
November 21 at 2h
December 21 at 0h
January 21 at 22h
February 21 at 20h

October 6 at 5h
November 6 at 3h
December 6 at 1h
January 6 at 23h
February 6 at 21h

EAST

WEST

NORTH

ZENITH

SEX

HYA

LEO

Regulus

CNC

M44

CMI

Procyon

MON

LYN

GEM

Pollux

Castor

M35

AUR

M37

M36

M38

Capella

Sirius

M41

CMA

Betelgeuse

ORI

M42

LEP

Rigel

TAU

Aldebaran

PER

Algol

M34

ARI

TRI

ERI

Mira

PSC

CET

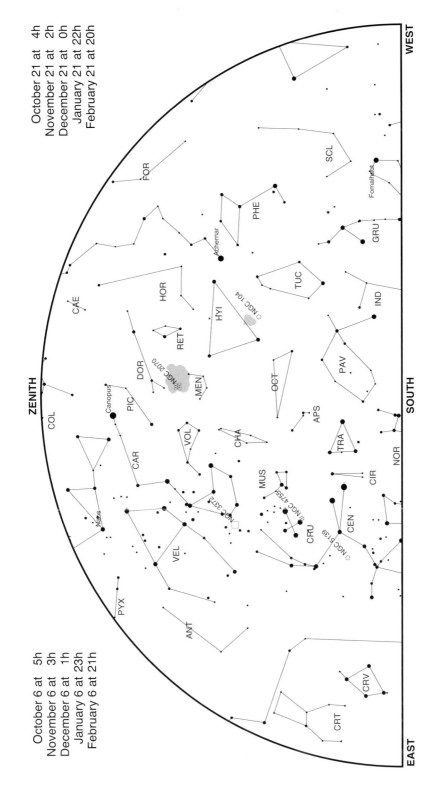

1S

October 21 at 4h
November 21 at 2h
December 21 at 0h
January 21 at 22h
February 21 at 20h

October 6 at 5h
November 6 at 3h
December 6 at 1h
January 6 at 23h
February 6 at 21h

WEST

ZENITH

SOUTH

EAST

SCL
FOR
PHE
Fomalhaut
GRU
Achernar
TUC
IND
HOR
CAE
HYI
NGC 104
RET
PAV
DOR
NGC 2070
MEN
OCT
COL
PIC
Canopus
CAR
CHA
APS
VOL
TRA
MUS
NOR
CEN
CIR
NGC 3372
NGC 4755
CRU
NGC 5139
Mars
VEL
PYX
ANT
CRT
CRV

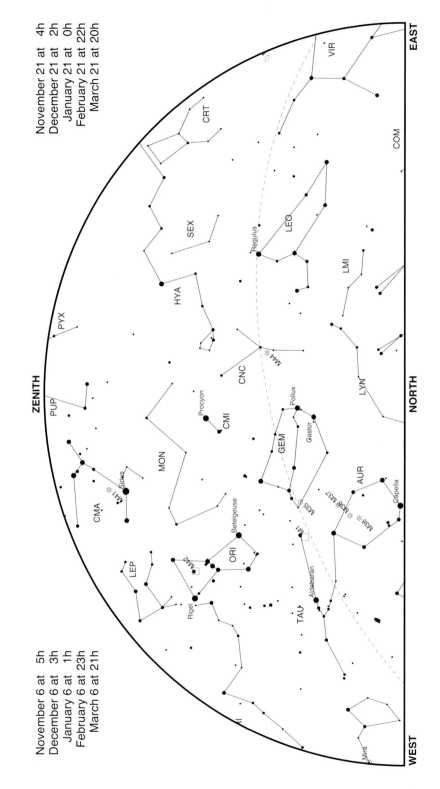

2N

ZENITH

EAST

NORTH

WEST

November 6 at 5h
December 6 at 3h
January 6 at 1h
February 6 at 23h
March 6 at 21h

VIR
CRT
COM
SEX
LEO
Regulus
LMI
HYA
PYX
CNC
M44
LYN
PUP
Procyon
CMI
MON
GEM
Pollux
Castor
AUR
Sirius
M41
CMA
M35
M36 M37
M38
Capella
Betelgeuse
M1
ORI
LEP
M42
Rigel
TAU
Aldebaran
Mira

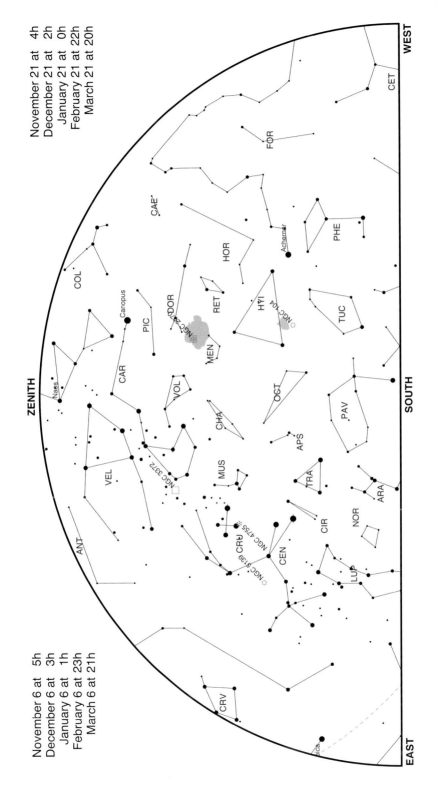

2S

WEST

November 21 at 4h
December 21 at 2h
January 21 at 0h
February 21 at 22h
March 21 at 20h

November 6 at 5h
December 6 at 3h
January 6 at 1h
February 6 at 23h
March 6 at 21h

ZENITH

SOUTH

EAST

CET

FOR

CAE

HOR

PHE

Achernar

RET

HYI

NGC 104

TUC

COL

DOR

NGC 2070

PIC

Canopus

MEN

VOL

CHA

OCT

PAV

APS

ARA

NOR

CIR

TRA

MUS

CEN

NGC 4755

CRU

NGC 3139

LUP

NGC 3372

VEL

ANT

CRV

Spica

Naos

3N

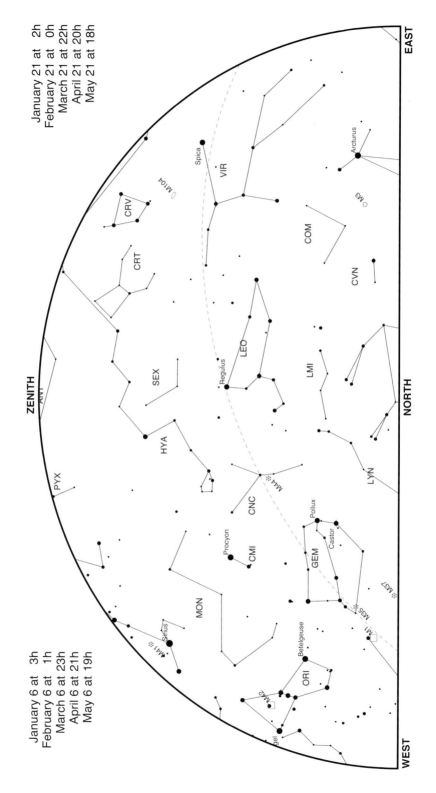

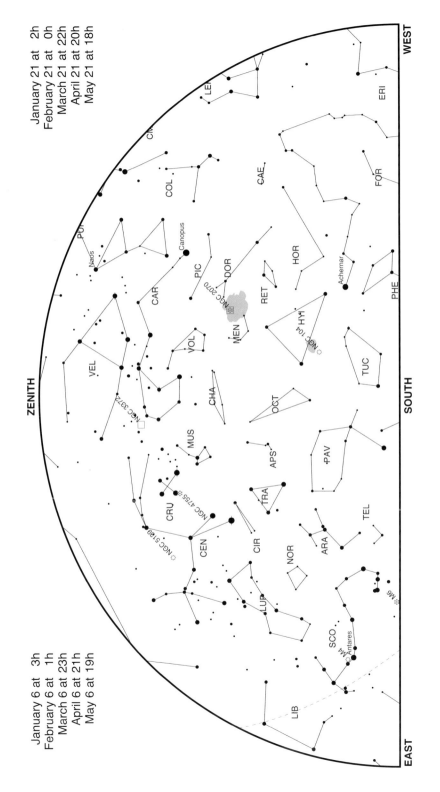

3S

WEST

EAST

SOUTH

ZENITH

January 21 at 2h
February 21 at 0h
March 21 at 22h
April 21 at 20h
May 21 at 18h

January 6 at 3h
February 6 at 1h
March 6 at 23h
April 6 at 21h
May 6 at 19h

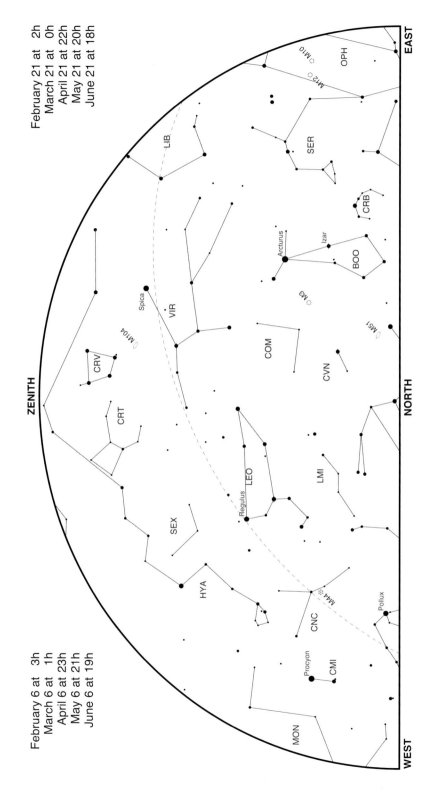

February 21 at 2h
March 21 at 0h
April 21 at 22h
May 21 at 20h
June 21 at 18h

February 6 at 3h
March 6 at 1h
April 6 at 23h
May 6 at 21h
June 6 at 19h

EAST

OPH

M10

M12

SER

LIB

CRB

Arcturus

Izar

BOO

Spica

VIR

M104

COM

M3

M51

CRV

CVN

ZENITH

CRT

LEO

LMI

NORTH

SEX

Regulus

HYA

CNC

M44

Pollux

Procyon

CMI

MON

WEST

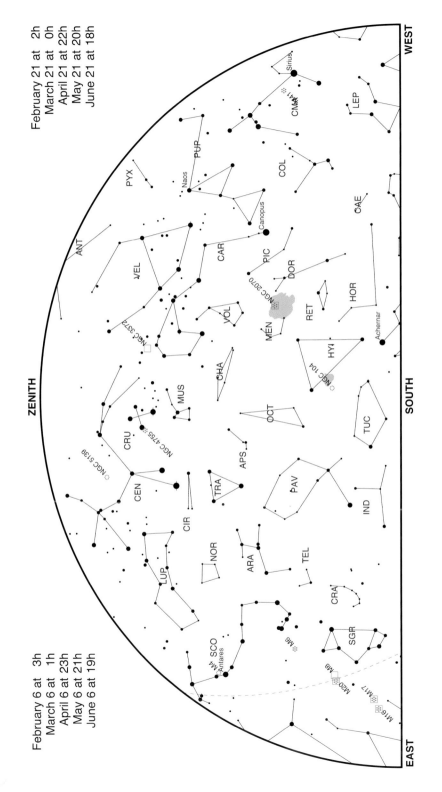

4S

WEST

ZENITH

SOUTH

EAST

February 21 at 2h
March 21 at 0h
April 21 at 22h
May 21 at 20h
June 21 at 18h

February 6 at 3h
March 6 at 1h
April 6 at 23h
May 6 at 21h
June 6 at 19h

Sirius
CMA
M41
LEP
PUP
PYX
Naos
COL
CAE
ANT
Canopus
CAR
VEL
PIC
DOR
NGC 3372
VOL
MEN
NGC 2070
RET
HOR
Achernar
CHA
HYI
NGC 104
MUS
OCT
CRU
APS
TUC
NGC 4755
NGC 5139
CEN
TRA
PAV
CIR
IND
NOR
ARA
TEL
LUP
CRA
SGR
SCO
M4
Antares
M6
M7
M20 M8
M16 M17

5N

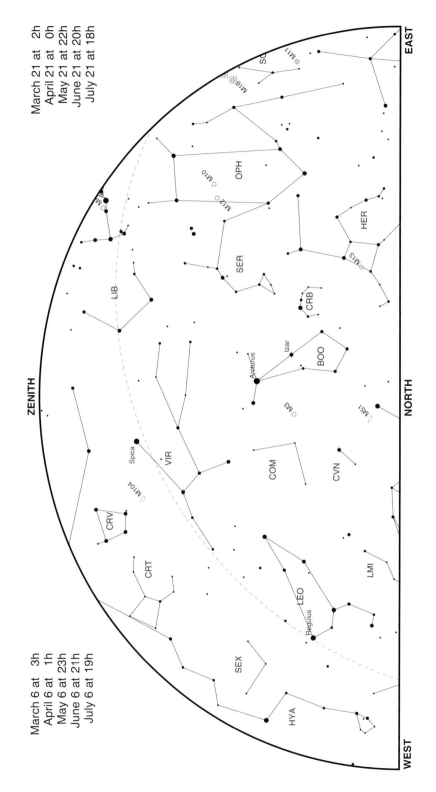

March 21 at 2h
April 21 at 0h
May 21 at 22h
June 21 at 20h
July 21 at 18h

March 6 at 3h
April 6 at 1h
May 6 at 23h
June 6 at 21h
July 6 at 19h

EAST

WEST

NORTH

ZENITH

SC
M16
M11
OPH
M10
M12
SER
HER
M13
CRB
LIB
Izar
BOO
Arcturus
M3
M51
Spica
VIR
COM
CVN
M104
CRV
CRT
LMI
LEO
Regulus
SEX
HYA

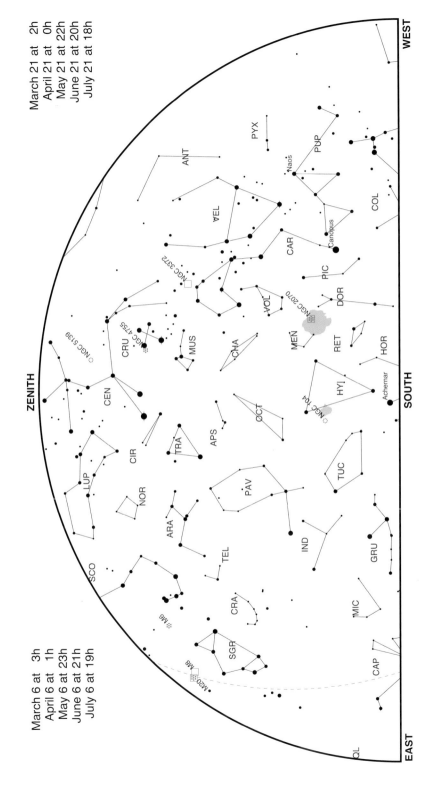

5S

March 21 at 2h
April 21 at 0h
May 21 at 22h
June 21 at 20h
July 21 at 18h

March 6 at 3h
April 6 at 1h
May 6 at 23h
June 6 at 21h
July 6 at 19h

WEST
ZENITH
EAST
SOUTH

PYX
ANT
PUP
Naos
VEL
COL
CAR
Canopus
NGC 3372
PIC
CRU
NGC 4755
VOL
DOR
NGC 5139
MUS
CHA
NGC 2070
CEN
MEN
RET
CIR
APS
OCT
HYI
Achernar
NGC 104
TRA
LUP
NOR
PAV
TUC
ARA
SCO
TEL
IND
GRU
M6
CRA
MIC
SGR
M20
M8
CAP

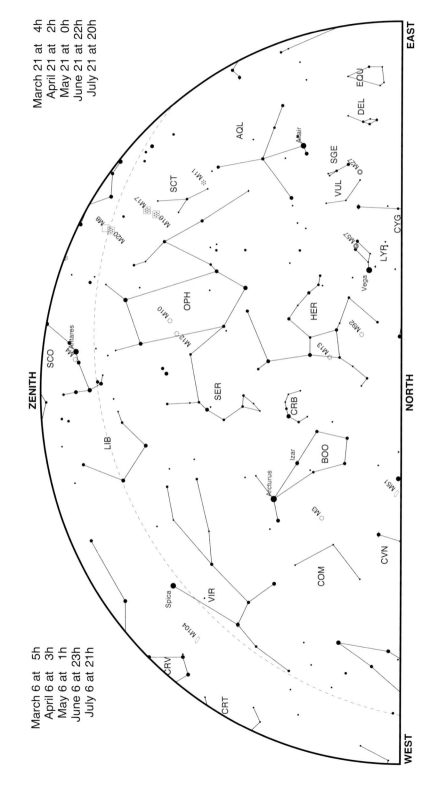

6N

EAST

ZENITH

NORTH

WEST

EQU
DEL
AQL
Altair
SGE
M27
VUL
SCT
M11
M17
M16
M20
M8
LYR
M57
Vega
HER
M92
OPH
M10
M12
M13
SCO
M4
Antares
SER
CRB
LIB
Izar
BOO
Arcturus
M3
M51
CVN
COM
VIR
Spica
M104
CRV
CRT
CYG

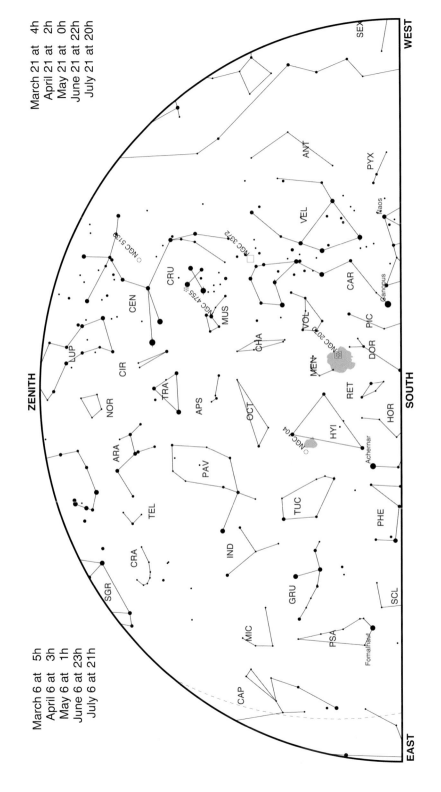

6S

March 6 at 5h
April 6 at 3h
May 6 at 1h
June 6 at 23h
July 6 at 21h

WEST

EAST

SOUTH

ZENITH

SEX

PYX

ANT

VEL

Naos

CAR

Canopus

PIC

DOR

RET

HOR

MEN

VOL

CHA

MUS

CRU

CEN

NGC 5139

NGC 3372

NGC 4755

NGC 2070

CIR

LUP

NOR

TRA

APS

OCT

ARA

TEL

PAV

IND

TUC

HYI

Achernar

NGC 104

PHE

SCL

GRU

MIC

PSA

Fomalhaut

CRA

SGR

CAP

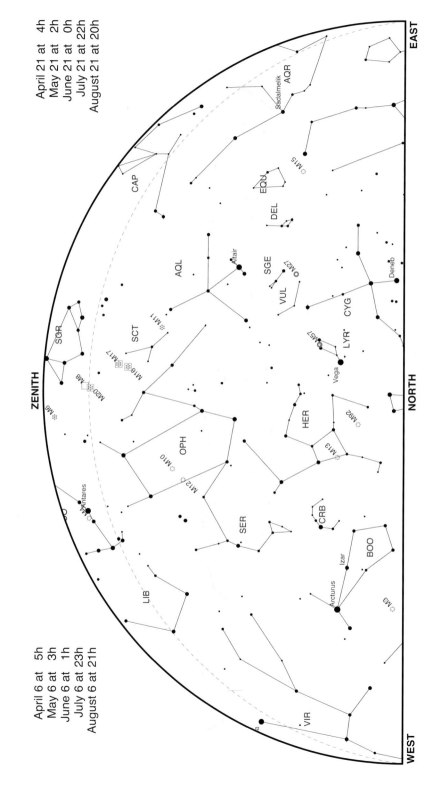

April 21 at 4h
May 21 at 2h
June 21 at 0h
July 21 at 22h
August 21 at 20h

April 6 at 5h
May 6 at 3h
June 6 at 1h
July 6 at 23h
August 6 at 21h

EAST

WEST

NORTH

ZENITH

AQR
Sadalmelik
M15
EQU
DEL
CAP
AQL
Altair
SGE
M27
VUL
CYG
Deneb
LYR
M57
Vega
HER
M13
M92
M11
SCT
M17
M16
SGR
M18
M20
M8
M6
OPH
M10
M12
SER
CRB
BOO
Izar
Arcturus
M3
VIR
LIB
Antares

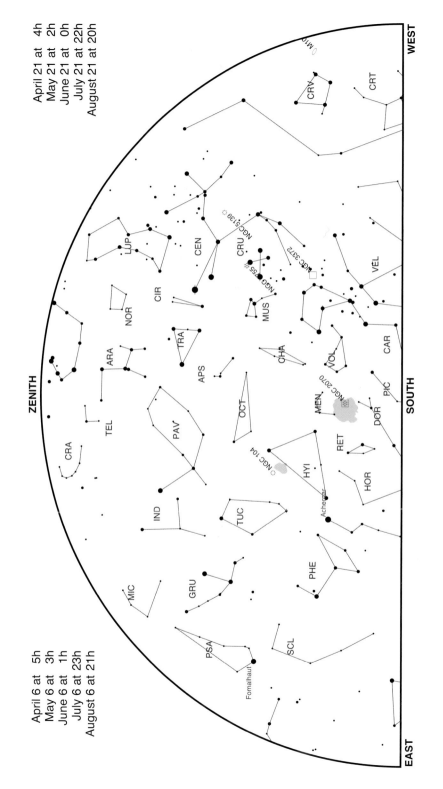

7S

WEST

EAST

SOUTH

ZENITH

April 21 at 4h
May 21 at 2h
June 21 at 0h
July 21 at 22h
August 21 at 20h

April 6 at 5h
May 6 at 3h
June 6 at 1h
July 6 at 23h
August 6 at 21h

CRV
CRT
ωMIC

LUP
CEN
CRU
NGC 5139
NGC 3372
NGC 755
MUS
VEL

CIR
NOR
TRA
APS
CHA
VOL
CAR

ARA
OCT
MEN
NGC 2070
PIC

CRA
TEL
PAV
HYI
RET
DOR

IND
TUC
Achernar
HOR

MIC
GRU
PHE

PSA
SCL
Fomalhaut

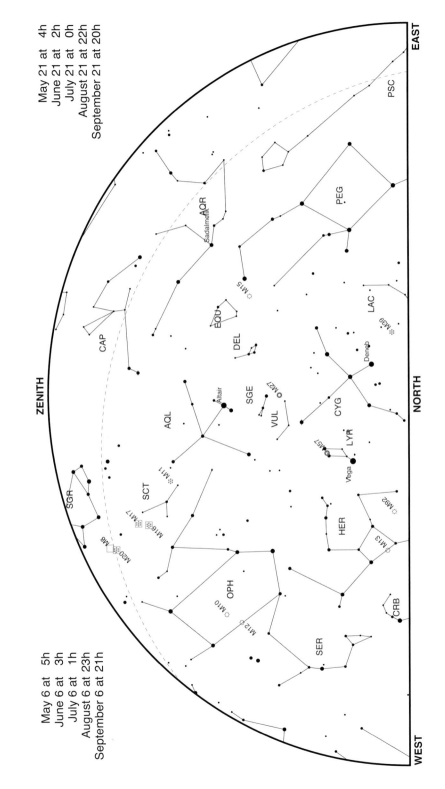

8N

May 6 at 5h
June 6 at 3h
July 6 at 1h
August 6 at 23h
September 6 at 21h

ZENITH

EAST

NORTH

WEST

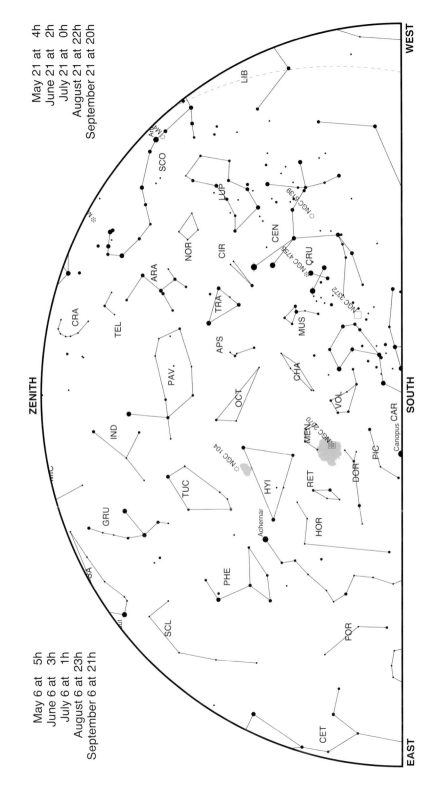

WEST

8S

May 21 at 4h
June 21 at 2h
July 21 at 0h
August 21 at 22h
September 21 at 20h

ZENITH

LIB

SCO
LUP

NGC 5139

CEN
NOR
CIR

ARA
NGC 4756
CRU

TRA
NGC 3372

CRA
APS
MUS

TEL

CHA

PAV.

VOL

OCT
SOUTH

IND
MEN

NGC 2070

Canopus CAR

TUC
NGC 104
HYI
RET
PIC

GRU
Achernar
DOR

HOR

PHE

SCL
FOR

CET

EAST

May 6 at 5h
June 6 at 3h
July 6 at 1h
August 6 at 23h
September 6 at 21h

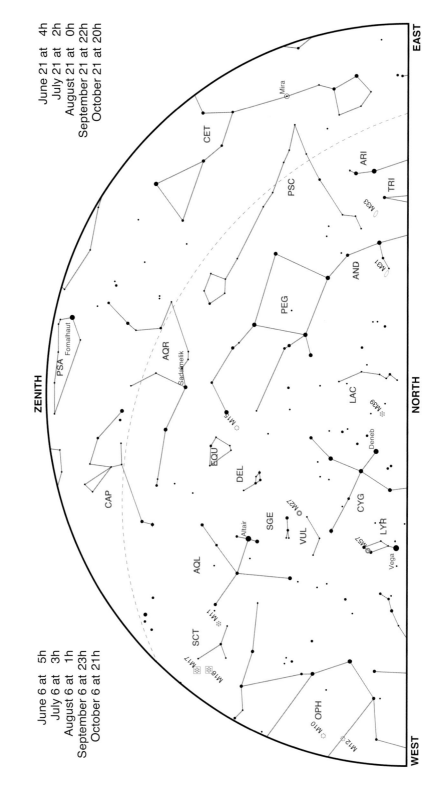

9N

June 6 at 5h
July 6 at 3h
August 6 at 1h
September 6 at 23h
October 6 at 21h

ZENITH

EAST

NORTH

WEST

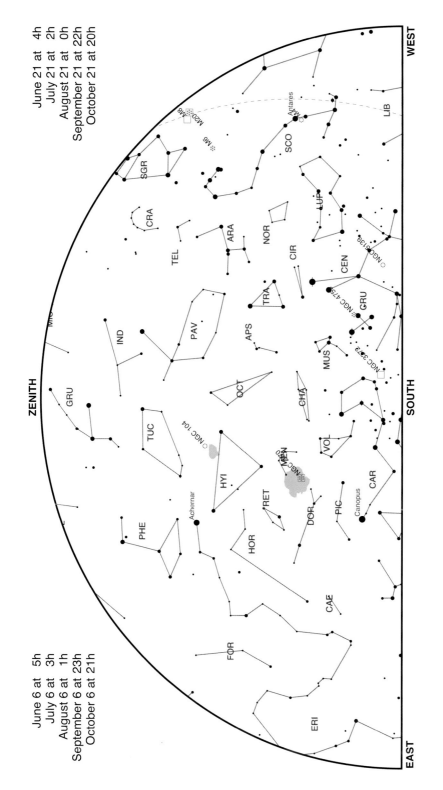

9S

June 21 at 4h
July 21 at 2h
August 21 at 0h
September 21 at 22h
October 21 at 20h

June 6 at 5h
July 6 at 3h
August 6 at 1h
September 6 at 23h
October 6 at 21h

ZENITH

SOUTH

EAST

M20 M8
M6
Antares
SCO
LIB
SGR
CRA
LUP
NGC6139
TEL
ARA
NOR
CIR
CEN
NGC4755
GRU
NGC3072
IND
PAV
TRA
APS
MUS
OCT
CHA
VOL
GRU
TUC
NGC 104
HYI
RET
NGC 2070
DOR
CAR
PHE
Achernar
HOR
PIC
Canopus
CAE
FOR
ERI

10N

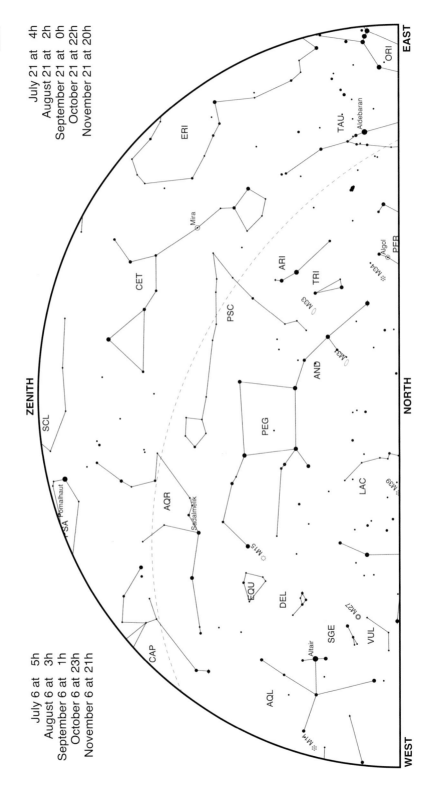

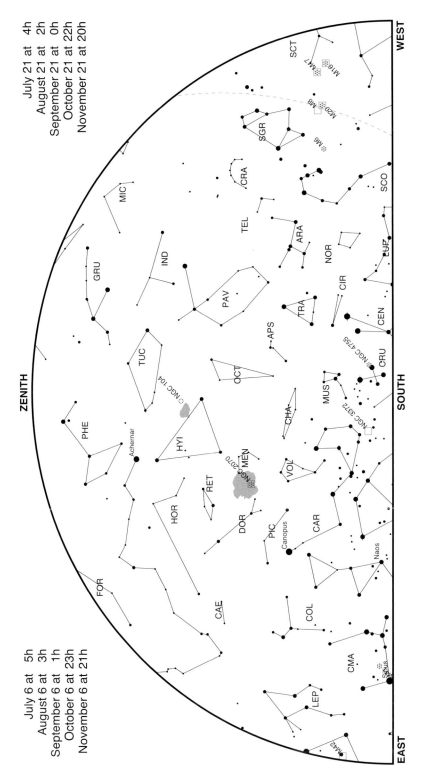

10S

WEST

ZENITH

SCT
M16
M20 M8
SGR
M6
CRA
SCO
MIC
TEL
ARA
NOR
IND
GRU
PAV
CIR
LUP
TRA
APS
CEN
TUC
NGC 104
OCT
CRU
NGC 4755
PHE
HYI
CHA
MUS
NGC 3372
Achernar
MEN
NGC 2070
VOL
RET
HOR
DOR
PIC
CAR
FOR
Canopus
Naos
CAE
COL
CMA
LEP
Sirius
M42

SOUTH

EAST

11N

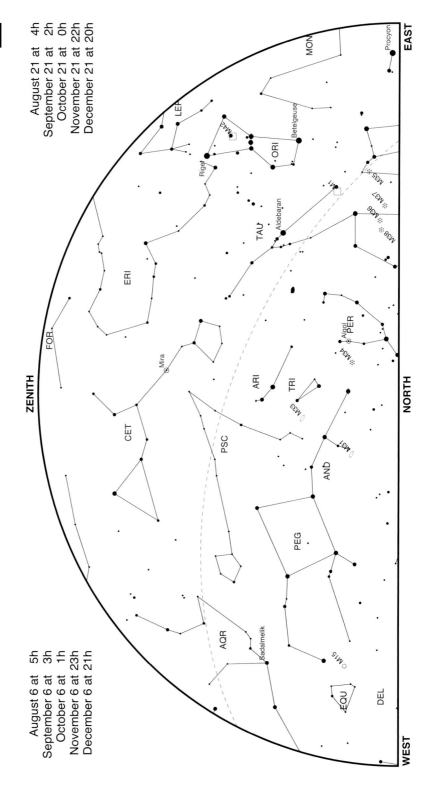

August 6 at 5h
September 6 at 3h
October 6 at 1h
November 6 at 23h
December 6 at 21h

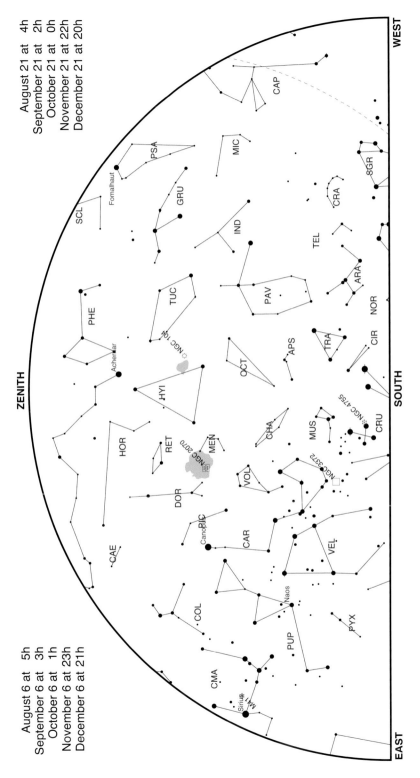

11S

WEST

August 21 at 4h
September 21 at 2h
October 21 at 0h
November 21 at 22h
December 21 at 20h

August 6 at 5h
September 6 at 3h
October 6 at 1h
November 6 at 23h
December 6 at 21h

EAST

SOUTH

ZENITH

CAP
PSA
MIC
SGR
Fomalhaut
GRU
CRA
SCL
IND
TEL
PHE
TUC
PAV
ARA
Achernar
NGC 104
NOR
HYI
APS
CIR
OCT
TRA
HOR
CHA
NGC 4755
RET
MEN
MUS
CRU
NGC 2070
DOR
VOL
NGC 3372
CAE
BIC
CAR
VEL
Canopus
COL
PUP
PYX
CMA
Naos
M41
Sirius

12N

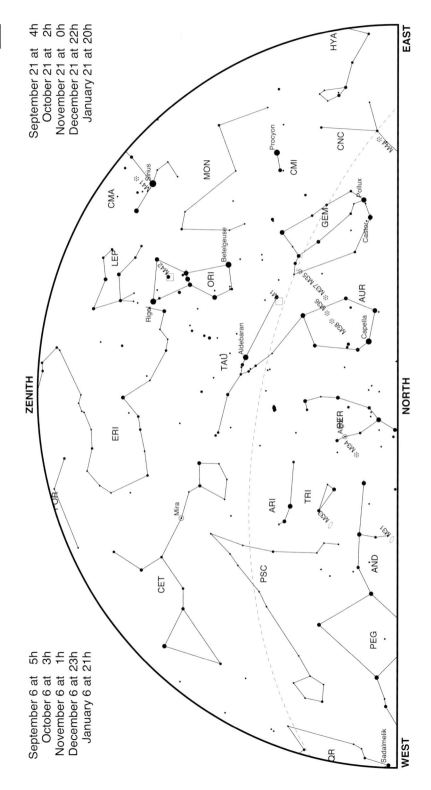

September 21 at 4h
October 21 at 2h
November 21 at 0h
December 21 at 22h
January 21 at 20h

September 6 at 5h
October 6 at 3h
November 6 at 1h
December 6 at 23h
January 6 at 21h

EAST

HYA

CNC

M44

Procyon

CMI

MON

Pollux

GEM

Castor

Sirius

M41

CMA

M35

M37

AUR

LEP

M42

Capella

ORI

M36

M38

Betelgeuse

Rigel

ZENITH

M1

Aldebaran

TAU

NORTH

ERI

PER

M34

FOR

PLE

Mira

AND

M33

TRI

ARI

M31

PSC

CET

PEG

Sadalmelik

AQR

WEST

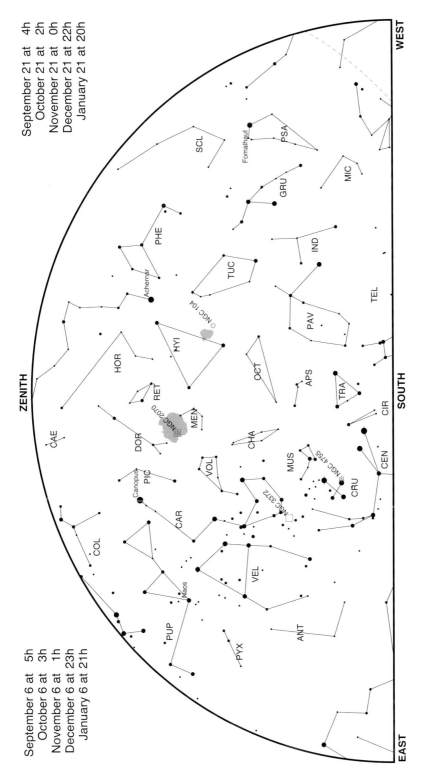

12S

WEST

ZENITH

SCL

PSA

Fomalhaut

GRU

MIC

PHE

Achernar

TUC

IND

NGC 104

HYI

PAV

TEL

HOR

RET

OCT

APS

TRA

CAE

DOR

NGC 2070

MEN

CHA

CIR

SOUTH

PIC

Canopus

VOL

MUS

NGC 4755

CEN

COL

CAR

NGC 3372

CRU

Maos

VEL

PUP

PYX

ANT

EAST

Phases of the Moon in 2018

Month	First Quarter	Full Moon	Last Quarter	New Moon
January	24	2 and 31	8	17
February	23	-	7	15
March	24	2 and 31	9	17
April	22	30	8	16
May	22	29	8	15
June	20	28	6	13
July	19	27	6	13
August	18	26	4	11
September	17	25	3	9
October	16	24	2 and 31	9
November	15	23	30	7
December	15	22	29	7

Eclipses in 2018

On 21 January, there will be a total eclipse of the Moon visible from North America (except eastern parts), Russia, Asia and northern Scandinavia. The eclipse begins at 10:50 UT and ends at 16:10 UT. The total phase begins at 12:31 UT and ends at 14:08 UT.

There will be a partial eclipse of the Sun on 15 February visible from most of Antarctica and southern South America. The eclipse begins at 18:56 UT and ends at 22:47 UT.

There will be a partial eclipse of the Sun on 13 July visible from Antarctica and south eastern Australia (including Tasmania). The eclipse begins at 01:48 UT and ends at 04:14 UT.

There will be a total lunar eclipse on 27 July visible from Antarctica, Australasia, Asia, Russia (except the north), Africa, Europe and East South America. The eclipse begins at 17:13 UT and ends at 23:30 UT. The total phase begins at 19:30 UT and ends at 21:14 UT.

There will be a partial eclipse of the Sun on 11 August visible from northern Canada, Greenland, Iceland, the Shetland Islands (the eclipse should *just* be noticeable in Lerwick), Scandinavia, most of Russia, most of Kazakhstan, Mongolia and most of China. The eclipse begins at 08:02 UT and ends at 11:31 UT.

Monthly Sky Notes and Articles

January

Full Moon: 2 January
New Moon: 17 January
Full Moon: 31 January

MERCURY starts the year at greatest elongation west, but it is barely visible in the dawn sky from the northern hemisphere because the ecliptic is so close to the horizon. It is better seen from the tropics and southern latitudes. However, its altitude is decreasing and it is lost in the morning twilight before the end of the month. It passes close to Saturn on 13 January and reaches aphelion on 25 January.

VENUS is lost to view this month, hidden in the solar glare as it undergoes superior conjunction on 9 January. Passing from morning to evening skies, it remains too close to the Sun to observe until well into February. Venus attains the first of two aphelions this year on 23 January.

EARTH reaches perihelion on 3 January, when its distance from the Sun is 147.1 million kilometres.

MARS is found in the morning sky this month. It is over $10°$ south of the celestial equator, giving good views to observers in tropical and southern latitudes but unfortunately staying quite low to the horizon when observed from farther north. It is found in the vicinity of Jupiter, passing within $0.2°$ of the larger, brighter planet on 7 January. Four days later, the waning crescent Moon forms a picturesque trio with Mars and Jupiter. Beginning the year at magnitude $+1.5$ in Libra, Mars passes into Scorpius on 31 January.

1 CERES is at opposition on the last day of the month and thus is visible most of the night. Binoculars or a small telescope are required to see this $+6.9$ magnitude dwarf planet in the constellation of Cancer.

JUPITER spends much of 2018 in the faint constellation of Libra. In January it is a morning sky object, beginning at magnitude –1.8 and brightening throughout the month. Only nearby Mars rivals it in brilliance. On 7 January, Mars and Jupiter approach within 0.2° of each other in an event that can be observed with the naked eye, binoculars or a small telescope. Then, on 11 January, the waning crescent Moon joins the pair. It passes close to Jupiter first, at approximately 06:00 UT, and then sails by Mars about four hours later. The separation of the objects will be too large to see the event through a telescope but the double appulse will be visible to the naked eye.

SATURN was at conjunction late last month and remains very close to the Sun at the beginning of this year. It is located in Sagittarius, well south of the celestial equator, making it difficult to see from northern temperate latitudes. However, observers in tropical and southern latitudes may be able to see the ringed planet low in the south east just before sunrise by the end of the month. It passes within 0.6° of Mercury on 13 January as the two planets cross paths in the sky.

URANUS is found in Pisces. It reaches its stationary point on 2 January and resumes direct motion. The planet is barely visible to the naked eye at magnitude +5.8, but is easily seen in binoculars. Uranus is at east quadrature on 14 January.

NEPTUNE is an evening sky object. It remains in Aquarius all year, its brightness hovering around eighth magnitude, making binoculars or a small telescope necessary in order to observe it. On 20 January, Neptune can be found 1.6° north of the waxing crescent Moon.

The Double 'Blue Moon' of 2018

David Harper

The second Full Moon in a calendar month is popularly known as a 'Blue Moon'. In 2018, there are two, since January and March have two Full Moons each.

The expression 'Blue Moon' has a long history in the English language. In Shakespeare's time, it denoted an evident absurdity or an event that would never happen, and by the 19th century it had taken on its more familiar meaning: a rare event, which happens 'once in a Blue Moon'.

Most ancient calendars were based directly on the phases of the Moon. Each month began at New Moon, so the Full Moon always fell in the middle of the month. This is still the case with the Islamic, Jewish and Hindu religious calendars and the traditional Chinese calendar. The Gregorian calendar, which is now the civil calendar for most of the world, has its origins in the calendar of

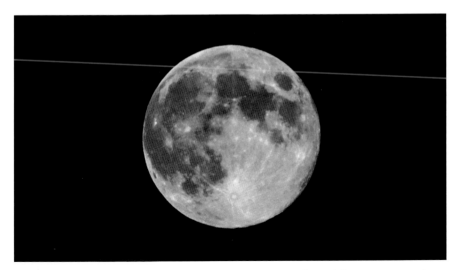

The Blue Moon of 31 August 2012 imaged by Cezar Popescu from Slobozia, Romania (Cezar Popescu http://atelierelealbe.eu)

ancient Rome. That was a lunar calendar in the earliest times, and it retained some lunar-like features such as 29-day months and a year of 355 days until Julius Caesar reformed it by setting the lengths of the months to the values that we still use today: 30 or 31 days, with 28 or 29 in February. Since most calendar months are longer than a lunar month, it is now possible for any month except February to contain two Full Moons.

The modern connection between Blue Moons and the calendar comes from American farmers' almanacs of the 19th century, which gave traditional names to the Full Moons in each season, such as the Harvest Moon and Hunter's Moon in autumn. There are normally three Full Moons per season, so when a season occasionally contained four Full Moons, the almanacs denoted the third one by the special name 'Blue Moon'. An article in a 1946 issue of Sky and Telescope magazine mistakenly interpreted the farmers' almanac usage to mean the second Full Moon in a calendar month, and when this appeared as a question in the board game Trivial Pursuit, it became established in popular culture.

This kind of Blue Moon happens once every three years on average, because 19 calendar years are almost exactly the same length as 235 lunar months. This is the well-known Metonic cycle. But 19 years contain 228 calendar months, and since these must contain 235 Full Moons, it is inevitable that 7 months in each Metonic cycle will have two Full Moons.

Sometimes, these Blue Moon months occur in quick succession. The most common case is a January/March pair, but January/April, January/May and December/March pairs are also possible. There was a January/March double Blue Moon in 1999, there is one this year (2018), and there will be another in 2037. The 19-year Metonic cycle appears in the sequence of double Blue Moons, but 1980 only had a single Blue Moon, and so will 2056, so simply counting every 19 years is not a foolproof way to predict when a double Blue Moon will occur.

However, 1999, 2018 and 2037 are part of a series of fourteen double Blue Moons separated by multiples of 19 years which began in 1695 (actually a December 1694/March 1695 pair) and will end in 2113. Another series began in 1915, followed by a further sixteen double Blue Moons ending in 2485. A third series begins with 2287 (a December 2286/March 2287 pair), continuing until 2857 and containing seventeen double Blue Moons. Taken together, these three series span almost twelve centuries and contain 48 double Blue Moons, giving an average of one double Blue Moon every 24 years. The longest gap between successive double Blue Moons during this period is 38 years, which happened between 1809 and 1847 and again between 1961 and 1999. Perhaps rare events should now be called 'once in a double Blue Moon'!

February

New Moon: 15 February

MERCURY is at superior conjunction on 17 February, moving from the morning sky to the evening. However, it is too close to the Sun to view most of this month so its appulse with Neptune on 25 February will likely go unobserved.

VENUS moves into the evening sky but remains a difficult object to spot near the horizon. On 21 February, it passes 0.6° south of Neptune but even bright Venus shining at magnitude –3.9 will be hard to see.

MARS moves from Scorpius to the non-zodiacal constellation of Ophiuchus on 8 February. The next day, it is visible several degrees south of the waning crescent Moon. It makes a pretty pairing with the first-magnitude star Antares on 11 February. The two objects are a similar reddish colour but Mars is slightly fainter. Southern hemisphere observers see Mars rising shortly after midnight, with the red planet well up in the east before dawn intrudes. From northern temperate latitudes, however, Mars remains quite low in the south east.

JUPITER is a bright magnitude –2.0 object in the faint constellation of Libra, rising after midnight. Never high in the sky when viewed from northern temperate latitudes, it gains useful altitude for observing the farther south you go. On 7 February, the waning crescent Moon passes near Jupiter. Quadrature on 10 February provides excellent astrophotographic opportunities as the shadows of the planet and its moons are cast slightly to one side.

SATURN rises before dawn in Sagittarius and is best seen south of the equator. It is 2.5° south of the waning crescent Moon on 11 February.

URANUS is found in the evening sky in Pisces and sets before midnight by the end of the month. On 20 February, the waxing crescent Moon passes within a few degrees of the faint planet; binoculars will be necessary to see the event.

NEPTUNE is becoming increasingly difficult to see low in the west after sunset. It has close encounters with both Mercury and Venus near the end of the month but by this time, it is in close proximity to the Sun and impossible to observe.

The James Webb Space Telescope

Richard Pearson

The James Webb Space Telescope (JWST) is set to launch in October 2018 aboard an Ariane 5 launch vehicle from French Kourou in NE Africa. The JWST is the successor to the Hubble Space Telescope although, unlike Hubble, the James Webb Space Telescope will not be in orbit around the Earth, but will travel around the Sun 1.5 million kilometres (1 million miles) away from the Earth at L2. This is one of the five so-called Lagrange Points, locations in space at which stable configurations in which three bodies can orbit each other (yet stay in the same positions relative to each other) can occur.

In order to carry out observations, JWST must be kept very cold. Otherwise infrared radiation from the telescope itself would overwhelm its instruments. To combat this, JWST has a large sunshield, rather like a huge parasol and measuring around 21 metres x 14 metres, which will allow the telescope to cool down to a working temperature of below –223 °C (50 Kelvin) by radiating its heat into space.

The James Webb Space Telescope carries four onboard instruments:

Near Infrared Camera (NIRCam) – an infrared imager that will have a spectral coverage ranging from the edge of the visible light through the near infrared.

Near Infrared Spectrograph (NIRSpec) – will also perform spectroscopy over the same wavelength range.

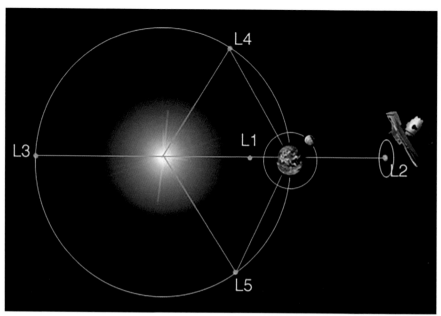

Diagram showing the Sun, Earth, JWST and the five Lagrange Points. Named after the 18th century French mathematician Joseph-Louis Lagrange, these are areas where gravity from the Sun and Earth balance the orbital motion of a satellite. These (highly-convenient) accidents of gravity and orbital mechanics are ideal places to park satellites. Other missions that have occupied Lagrange Points include the Solar and Heliospheric Observatory (SOHO) at L1 and the Herschel Space Observatory at L2 (JWST/NASA)

Mid-Infrared Instrument (MIRI) – will measure the mid-infrared wavelength range from five to 27 micrometers. Containing both a mid-infrared camera and an imaging spectrometer MIRI will work at a temperature of –266°C (7 Kelvin) using a helium refrigerator, or cryocooler system

Fine Guidance Sensor; Near Infrared Imager, and Slit-less Spectrograph (FGS/NIRISS) – will be used to stabilize the line-of-sight of the observatory during science observations. Measurements by the FGS are used both to control the overall orientation of the spacecraft and to drive the fine steering mirror for image stabilization. The near-infrared instruments NIRCam, NIRSpec and FGS/NIRISS will operate at around –234°C (39 Kelvin) through a passive cooling system.

As big as tennis court and as tall as a four-story building, a full-scale model of the James Webb Space Telescope was on display from 8 to 10 March 2013 at the South by Southwest Interactive Festival in Austin, Texas. NASA's James Webb Space Telescope is the successor to Hubble and the largest space telescope to ever be built. The James Webb Space Telescope is named after James Edwin Webb (1906 -1992), NASA's second administrator from February 1961 to October 1968 (NASA / Chris Gunn)

JWST has sufficient onboard fuel to keep the telescope operational for 10 years, although because it will be too far away from Earth for NASA to launch a manned repair mission in 2028, it will simply become non-operational. However, NASA is now studying plans to launch a Hubble Space Telescope repair mission to extend HST's lifespan, thereby enabling astronomers to continue imaging the universe long after the JWST ceases to operate.

Equipped with a 6.5 metre diameter primary mirror, JWST is the largest space telescope ever built and will be invaluable in helping us to scrutinize the early Universe.

The Universe came into existence some 13.8 billion years ago in an event called the Big Bang and during the following 9 billion years the first stars formed out of the abundant hydrogen gas clouds. These stars were immense and, as they went through their evolutionary processes, the nuclear fusion processes inside their cores gave rise to heavier and heavier elements. Their huge size doomed them to relatively short life spans of a few million years before eventually exploding as violent supernova. We are all made from star stuff due to our Sun and solar system having evolved from the remnant of a supernova created by one of the first stars to explode around 4.6 billion years ago.

The first protogalaxies (masses of gas from which galaxies are thought to have developed) then formed as the universe continued to expand, ultimately forming a cosmic web of galaxy clusters throughout the Universe. Each galaxy, including our own Milky Way, consists of billions of stars, the light from the smallest, most red-shifted galaxies originating nearly 200 million years after the Big Bang.

The distance from Earth to the edge of the observable universe is around 45 billion light years in any direction. The James Webb Space telescope will take images and rainbow spectra of this early period of time to allow astronomers to learn much about the very first stars and galaxies in the early universe.

March

Full Moon: 2 March
New Moon: 17 March
Full Moon: 31 March

MERCURY emerges from the solar glare in the west after sunset. For observers in northern temperate latitudes, this is the best apparition of the year. On 4 March, magnitude –1.0 Mercury is found 1.1° north of the much brighter planet Venus. It reaches perihelion on 10 March and attains greatest elongation east five days later. It has another encounter with Venus on 19 March although not as close as close as the one earlier in the month. A stationary point is reached on 22 March when Mercury switches from direct to retrograde motion and the planet soon vanishes from view as it approaches conjunction next month.

VENUS continues to gain altitude after sunset, more quickly from northern hemisphere viewpoints than elsewhere on Earth. On 4 March, it is less than two degrees from its fellow inferior planet Mercury. At magnitude –3.9, it outshines Mercury by almost three magnitudes. It is 3.7° north of the day-old Moon at approximately 20:00 UT on 18 March and 3.8° south of Mercury on the following day. Venus has another planetary encounter on 29 March when it passes only 0.1° south of Uranus. Binoculars or a telescope will be necessary to see both Venus and faint Uranus, shining at magnitude +5.9, low in the darkening sky, although this will be difficult to observe due to the proximity to the Sun of both planets.

EARTH reaches equinox on 20 March. On this day, the Sun crosses the celestial equator, passing from south to north, and heralding astronomical autumn in the southern hemisphere and astronomical spring in the north. The Earth will pass through a second equinox in September.

MARS appears near the waning crescent Moon on 10 March. The following day, the red planet moves from the non-zodiacal constellation of Ophiuchus to Sagittarius. West quadrature is reached on 24 March. This is when the angle between Mars and the Sun is 90° as seen from the Earth. Mars may be found south of the ecliptic between Jupiter and Saturn in the morning sky. Southern viewers get much the best views, with Mars rising around midnight to appear high in the east before the skies lighten at dawn. Mars brightens by half a magnitude this month, from +0.8 to +0.3

JUPITER now rises before or around midnight and is well-placed for observing in the early morning hours from southern and tropical latitudes. Unfortunately for those in northern temperate regions, Libra never rises very high above the horizon so Jupiter does not gain much altitude, even when transiting. It is south of the waning gibbous Moon on 7 March and two days later, reaches a stationary point, changing from direct to retrograde motion.

SATURN is a morning sky object, shining at magnitude +0.5, and best seen from the southern hemisphere where it can be found well aloft in the east during the early morning hours. On 11 March, the planet passes about 2° south of the waning crescent Moon. West quadrature occurs on 29 March. This is usually a good time to observe and photograph Saturn as the shadow of the disk of the planet is cast slightly to one side on the rings, rendering an interesting three-dimensional effect. Unfortunately for observers in northern temperate latitudes, Saturn remains fairly low to the horizon in Sagittarius.

URANUS is becoming increasingly difficult to observe in the evening sky as conjunction with the Sun approaches next month. Look for it in the constellation of Pisces as soon as the sky is dark. It sets by mid-evening early in the month. On 19 March, Uranus can be found around 5° north of the thin crescent Moon and on 29 March, it has a very close encounter with Venus. However, this will be difficult to observe due to both planets' proximity to the Sun.

NEPTUNE is at conjunction on 4 March and is too close to the Sun to see this month.

A Closer Look at Antlia and Pyxis:
The Pump and Compass

Brian Jones

Antlia

The tiny and rather obscure Antlia (the Air Pump) is one of the 14 constellations devised by the French astronomer Nicolas Louis de Lacaille during his stay at the Cape of Good Hope in 1751/52 (for further details see the article *Nicolas Louis De La Caille: Bringing Order to the Southern Skies* in the *Yearbook of Astronomy 2017*).

Lacaille created this group to commemorate the air pump invented by the French physicist Denis Papin. As with many of the constellations devised by Lacaille, both Pyxis and Antlia lie in an area of sky devoid of prominent stars. The chart shows Antlia together with the neighbouring Pyxis (see below), the bright star Naos in the adjoining constellation Puppis being included here as a guide to tracking down these two faint groups. Both Antlia and Pyxis can be seen from latitudes south of around 50°N and are therefore observable from

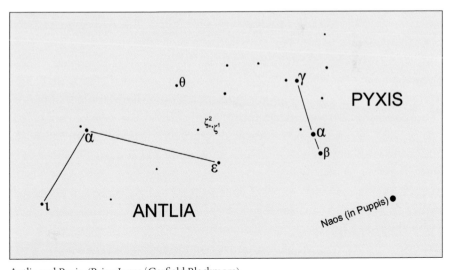

Antlia and Pyxis. (Brian Jones/Garfield Blackmore)

the northern United States, central Europe, northern China and from any latitude to the south of these.

The constellation takes the form of a bent line of three inconspicuous stars, the brightest being the magnitude 4.28 orange giant Alpha (α) Antliae, the light from which has taken over 350 years to reach us. A short way to the west of Alpha is Epsilon (ε) Antliae, another orange giant star which shines at magnitude 4.51 from a distance of around 700 light years.

The crooked line of stars forming the main part of this unimpressive constellation is completed by a third orange giant, this being Iota (ι) Antliae. Situated near the eastern border of Antlia, the magnitude 4.60 glow of Iota set off towards us around 190 years ago. The white giant Theta (θ) Antliae is the fourth brightest member of Antlia, the light from this magnitude 4.78 star having taken around 340 years to reach our planet.

A Double Star

The wide optical double star comprising **Zeta¹ (ζ^1)** and **Zeta² (ζ^2) Antliae** can be found just to the west of a line from Epsilon to Theta as shown on the chart. Shining at magnitude 5.76 from a distance of around 410 light years, Zeta¹ is marginally the brighter of the two. Magnitude 5.91 Zeta² is slightly nearer, its light having taken around 380 years to reach us.

Binoculars will easily resolve both components, although Zeta¹ is itself a binary star, and closer examination through a small telescope may reveal the seventh magnitude companion star which lies quite close by.

Pyxis

The equally small and obscure constellation Pyxis (the Mariner's Compass) appears as nothing more than a line of faint stars to the west of Antlia, and was created by Lacaille to represent the magnetic compass used by sailors and navigators, appropriately located as it is near the stern of the former constellation (and now-dismantled) Argo Navis.

Pride of place in this inconspicuous group goes to Alpha (α) Pyxidis, a magnitude 3.68 blue giant star shining from a distance of almost 900 light years. Lying immediately to the south of Alpha is Beta (β) Pyxidis, the magnitude 3.97 glow from this yellow giant star having taken a little over 400 years to reach us. The trio of stars denoting the main part of Pyxis is rounded off by Gamma (γ) Pyxidis, a magnitude 4.02 orange giant located at a distance of just over 200 light years.

April

New Moon: 16 April
Full Moon: 30 April

MERCURY is at inferior conjunction on the first day of the month and moves back into the morning sky for what is this year's best apparition of the planet for southern hemisphere observers. It rapidly gains altitude in the north east throughout the month, right through to greatest elongation west on 29 April. For those situated in northern temperate latitudes, Mercury remains fairly low to the horizon throughout the month. On 14 April, whilst a few degrees away from the waning crescent Moon, it reverses direction from retrograde to direct motion. Aphelion occurs on 23 April.

VENUS appears higher above the western horizon after sunset every evening, rising faster in the northern hemisphere than in the south, and remaining at magnitude of –3.9. On 17 April, the day-old Moon is within 6° of Venus.

MARS passes 1.3° south of Saturn on 2 April in an event visible to the naked eye or through binoculars. The waning gibbous Moon is found in the neighbourhood of Mars five days later. Mars is in Sagittarius so the view improves the farther south you go. In the southern hemisphere, the red planet rises before midnight and is high above the horizon during the morning hours.

JUPITER has two encounters with the Moon this month. On 3 April, it is found in close proximity with the waning gibbous Moon and on the last day of the month, is approximately 4° south of the Full Moon. Observers in tropical and southern temperate latitudes will get the best views as Jupiter, currently in Libra, is high in the sky and shining at magnitude –2.4

SATURN has a close encounter with the slightly brighter Mars on the second day of the month. On 7 April at 13:30 UT, it is found only 1.9° south of the waning gibbous Moon. Saturn is at aphelion on 17 April, an event which happens only every 29.5 years. It arrives at a stationary point the following day, reversing direction from direct to retrograde. Found in Sagittarius, this first-magnitude object is becoming ever easier to observe in the early morning hours as it distances itself from the Sun.

URANUS arrives at conjunction on 18 April, rendering the planet unobservable this month. Nine days later, Uranus passes from Pisces into Aries.

NEPTUNE was at conjunction early last month and is just emerging from morning twilight. Located in Aquarius, it is difficult to observe from northern temperate latitudes where the ecliptic is quite close to the horizon at this time of year. On 12 April, the waning crescent Moon passes within $2°$ of the blue ice giant.

Fate or Destiny:
Our Solar System as an Abode for Life

Richard Pearson

To the layperson it seems that many of the spacecraft that have been sent out to explore the solar system have detected organic molecules and water, the two main ingredients that support the origin of life, so we are most likely to find 'life' on Mars in the near future.

It was only during the last few years that NASA and European spacecraft have been equipped with instruments sensitive enough to detect both water and Deoxyribonucleic Acid (DNA). This is a molecule which carries the genetic instructions used in the growth, development, functioning and reproduction of all known living organisms and contains the biological instructions that make each species unique.

In April 1980, during a lecture at Cardiff University, astronomer Fred Hoyle made a bold statement that DNA could exist on comets, and that these nomads

Artist's impression of Giotto and Halley's Comet. (ESA)

HMC 68 Image Composite
Comet Halley   14th March 1986

© MPAe 1986,1990   23–AUG–1990 13:09

Halley's Comet as seen by ESA's Giotto space probe in 1986, the last time the comet visited the inner Solar System. Giotto was ESA's first deep space mission, and obtained the first close-up images of a comet. This image was taken from a distance of about 2000 km from the comet, the Sun being located towards the left of the image, provoking outbursts of gas and dust from the comet's nucleus. (ESA/MPS)

were capable of seeding life on Earth and other planets in our solar system.

In 1986 the celebrated Halley's Comet returned to the inner solar system and was met by a flotilla of spacecraft, including the European Space Agency Giotto probe whose achievements included being the first spacecraft to image the nucleus of a comet. During its encounter, the Giotto probe was pummelled with cometary grains and dust from the coma surrounding the nucleus of Halley's Comet. Images taken by Giotto showed Halley to be almost entirely black and scientists analyzing the particles striking the Giotto probe's forward shield determined that much of the material was organic in nature.

Water has been detected in craters in the polar regions of both Mercury and the Moon, its existence made possible due to the craters being in permanent shade from sunlight. Equally surprising is that scans and observations made by instruments on board NASA's Dawn space probe, which entered into orbit around the minor planet Ceres in March 2015, have revealed a 100km thick mantle containing up to 200 million cubic kilometres of water, which is more than the amount of fresh water on Earth.

Images from the Hubble Space Telescope taken in 2012 seemed to reveal columns of water vapour emanating from Europa's icy exterior. Subsequent images taken by Hubble in 2014 provided further evidence of their existence, appearing to show what could be 200km high plumes of water vapour gushing

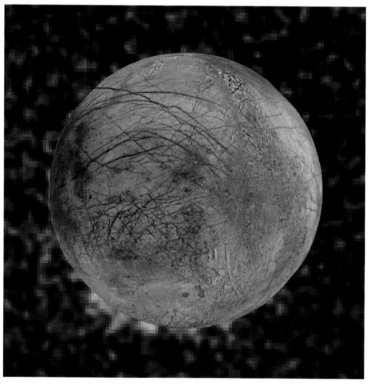

A composite image, obtained by NASA's Hubble's Space Telescope Imaging Spectrograph, shows suspected plumes of water vapour erupting from the limb of Jupiter's moon Europa and located here at the 7 o'clock position. The plumes, which are believed to come from a subsurface ocean on Europa, were seen in silhouette as the moon passed in front of Jupiter. Scientists observed what could be water plumes erupting from Europa on three of the 10 occasions during a 15-month period when the moon was seen passing in front of Jupiter. The Hubble data were taken on 26 January 2014. The image of Europa, superimposed on the Hubble data, is assembled from material from the Galileo and Voyager missions. (NASA/ESA/W. Sparks (STScI)/USGS Astrogeology Science Center)

from the Jovian moon. Data returned from NASA's Voyager and Galileo space probes suggested that Europa has a subsurface ocean, which many believe may contain life. NASA scientists consider that a layer of liquid water exists beneath Europa's surface, and that heat generated by tidal flexing due to the gravitational tug of Jupiter on the moon allows the ocean to remain liquid.

NASA's Cassini spacecraft has detected similar fountains on Saturn's moon Enceladus which release plumes of water vapour and chemicals high above the moon's surface. Planetary scientists believe that the water originates from a subsurface ocean. The presence of volcanic activity means that hydrothermal vents may exist on the ocean floor. Similar vents found on Earth are known to release chemicals that are a source of food for a multitude of sea creatures.

NASA's New Horizons spacecraft has provided firm evidence that the dwarf planet Pluto also has a subsurface ocean. In September 2016, scientists at Brown's University simulated the impact that is believed to have formed Sputnik Planitia, a highly-reflective ice-covered basin measuring around 1,050 kilometres (650 miles) by 800 kilometres (500 miles) which lies mainly in Pluto's northern hemisphere. The simulation showed that the basin may have been the result of liquid water

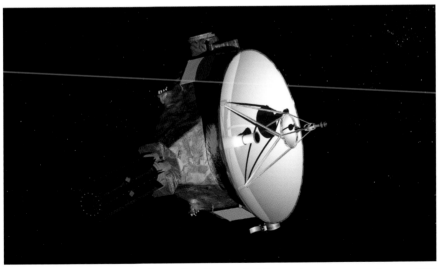

NASA's New Horizons spacecraft, seen here in an artist's impression en route to a January 2019 encounter with Kuiper Belt object 2014 MU69, provided scientists with evidence that the dwarf planet Pluto also has a subsurface ocean. (NASA/JHUAPL/SwRI)

up welling from below the surface after the collision, implying the existence of a subsurface ocean at least 100 kilometres (62 miles) deep.

The solar system as an abode for life is looking promising, and with new extremely large telescopes joining the search over the next five years, the chances of discovering life forms looks better than ever. Given the right conditions, if life can find a way to form and evolve it will – even if it differs to what we are used to here on Earth …

May

New Moon: 15 May
Full Moon: 29 May

MERCURY is best seen in the morning sky from the southern hemisphere, early in the month, before the planet gets too close to the horizon. On 12 May, it is only 2° south of Uranus and the following day it is a similar distance north of the waning crescent Moon. It vanishes into the dawn twilight late in the month, on its way to conjunction with the Sun in June.

VENUS reaches perihelion on 15 May and makes a pretty picture with the nearby waxing crescent Moon two days later. This evening apparition still favours the northern hemisphere but only for another month or so. Venus brightens slightly this month, from magnitude –3.9 to –4.0

MARS is well-placed for viewing in the morning sky for southern hemisphere observers. On 6 May, the waning gibbous Moon passes close by and on 14 May, Mars moves from Sagittarius into Capricornus.

JUPITER is at opposition on 9 May when it reaches a maximum magnitude of –2.5. It is visible all night in Libra but is best seen from southern temperate latitudes where the ecliptic rises high in the sky. The waxing gibbous Moon makes a close pass by Jupiter on 27 May.

SATURN is primarily a morning sky object and fairly low to the horizon for viewers in the northern hemisphere. However, it rises before midnight and is quite high in the sky for tropical and southern observers. On 4 May at 20:30 UT, it is 1.7° south of waning gibbous Moon in Sagittarius.

URANUS is located in Aries. It is still too close to the Sun to observe early in the month but for southern hemisphere viewers, it reappears low in the north east around mid-month. Observers in northern temperate latitudes will have to wait a little longer before Uranus appears low in the south east before dawn. On 12 May, Uranus and Mercury pass by each other although this event may be unobservable due to solar glare. The following day, the thin crescent Moon makes a close approach to Uranus.

NEPTUNE is best seen from southern latitudes. From the northern hemisphere is it very low in the south east at dawn and difficult to see. On 10 May it can be found north of the waning crescent Moon in Aquarius.

The Royal Greenwich Observatory

Richard Pearson

The Royal Greenwich Observatory was established in 1675 by King Charles II to provide the means of solving the longitude problem, and over the next 100 years it was the duty of the Astronomers Royal to precisely measure the positions of the stars from the observatory and to publish accurate star atlases. Even today, the Nautical Almanac is being produced long after the fifth Astronomer Royal, Nevil Maskelyne, first published it in 1766 (with data for 1767) as the *Nautical Almanac and Astronomical Ephemeris*.

In spite of each Astronomer Royal working hard to accurately record their observations, the Greenwich astronomers did not solve the longitude problem. This honour fell to the Yorkshire-born clock maker John Harrison who designed and constructed a number of clocks before inventing the marine chronometer, a timepiece which was precise and accurate enough to maintain Greenwich Mean Time over long voyages, revolutionising navigation and allowing mariners to safely sail the seas without getting lost or shipwrecked.

The Curator of the Observatory is Dr. Louise Devoy who I had the pleasure of meeting while filming my monthly programme Astronomy & Space. Louise's specialities are astrolabes, sundials and quadrants, examples of which are on show at the Observatory.

Flamsteed House showing the iconic red Time Ball on the roof of the building. The Time Ball was first used in 1833 and is still in operation. At 12.55 each day, the time ball rises half way up its mast and at 12.58 rises all the way to the top. At 13.00 exactly, the ball falls, and so provides a time signal to anyone who happens to be looking. (Robert Batey / Richard Pearson)

The octagonal two-story building was the original observatory on the site, and is named after the first Astronomer Royal John Flamsteed. The rooms on the ground floor were used both as living accommodation and as a place to entertain visitors. The observatory was funded by the Admiralties who were frequent visitors.

Soon after Edmund Halley became second Astronomer Royal in 1720, the Transit Room was built nearby at the side of the courtyard. The first instrument installed was Halley's 8-foot Iron Mural Quadrant, which was completed in 1725. When Halley arrived as Flamsteed's successor he found the Observatory to be empty of instruments. The Quadrant was commissioned as a replacement for Flamsteed's Mural Arc which had been removed, together with all the other instruments, by Margaret Flamsteed when she left the Observatory in early 1721 following the death of her husband.

As the years passed, two more transit instruments, each incorporating major improvements, were put in place, the first being a 10 cm (4-inch) aperture telescope set between two white posts, which was used by Halley's successor as Astronomer Royal, James Bradley.

Next in line is the grandest of all the transit instruments. Erected in 1850 by Sir George Biddle Airy, who became the seventh Astronomer Royal in 1835, it was first used in January 1851. The Airy Transit Circle has an aperture of 8 inches (20cm) and a telescope tube almost 12 feet (3.65 metres) in length. The instrument was used on every clear evening, Airy's assistants writing down the precise time as he called out the moment one of the stars passed due south over the Greenwich Meridian.

The upper room of Flamsteed House was by this time set out with rows of wooden tables where sat young men and women (known as computers) for long hours, carefully reducing the previous night's observations.

Dr Louise Devoy was kind enough to show me the two prized telescopes now on display at the Observatory. Behind the Transit Room is a pleasant lawn surrounded by an area planted with shrubs, and it is here that the remaining 10 feet (3 metre) section from the mirror end of Sir William Herschel's Great Forty-Foot telescope is on display. Originally constructed between 1785 and 1789 and set up in the grounds of Observatory House, William Herschel's

The author talking to Dr. Louise Devoy while standing in front of the bottom section of Sir William Herschel's Forty-Foot telescope. Inset is a depiction of Sir William Herschel. (Robert Batey / Richard Pearson)

The 28-inch refractor telescope inside the magnificent onion dome of the Royal Greenwich Observatory. (Robert Batey/Richard Pearson)

home in Slough, the Forty-Foot was a reflecting telescope with a 48 inch (122cm) diameter primary mirror. Sadly, because of the poor reflectivity of the speculum-metal mirrors of the time, it was seldom used for observations.

Sir George Biddle Airy was succeeded by the eighth Astronomer Royal, Sir William Christie, in 1881. By this time things were changing, astronomers now being more interested in recording the spectra of stars and measuring the separations of double stars. Christie was influential in persuading the Admiralty to pay for the construction of a 28 inch (71cm) refractor. Ordered in 1885 and brought into use in 1893, the 28-inch Greenwich refracting telescope is the largest of its type in the UK and the seventh largest in the world. Its unusual design meant that it was suitable for both visual and photographic use. Constructed by Sir Howard Grubb of Dublin, the telescope was a replacement for the smaller 12.8 inch (32.5cm) Merz Refractor which originally occupied the South-East Dome (now known as the Great Equatorial Building) and on whose mounting it was placed.

The Greenwich refracting telescope is a wonderful instrument which I know very well. With Dr Louise Devoy at my side, we stood close to the eyepiece and looked up at the magnificent sight presented by the matrix tube towering above us. The telescope was primarily used for measuring double stars, and was seldom used for astrophotography.

Looking over Greenwich Park from the observatory, the view is breath taking, encompassing the National Maritime Museum and Old Royal Naval College, the curve of the River Thames, St Paul's Cathedral, the London Eye and much more. When architect Sir Christopher Wren chose this site for the Royal Greenwich Observatory, he truly put it at the heart of the world, while the Greenwich Meridian is now known worldwide.

June

New Moon: 13 June
Full Moon: 28 June

MERCURY is too close to the sun to be observed early in the month as it passes through superior conjunction on 6 June. It coincidentally reaches perihelion on the same day. It reappears in western skies by mid-month; this evening apparition is almost equally favourable for northern and southern hemispheres. On 24 June, Mercury passes 5° south of Pollux, the brightest star in the constellation of Gemini.

VENUS continues to slowly brighten in the evening sky where the planet now surpasses magnitude –4.0. On 8 June it passes a few degrees south of Pollux. The waxing gibbous Moon glides past Venus even more closely on 16 June.

EARTH is at solstice on 21 June. This is the day that the Sun reaches its most northerly declination. It is the beginning of astronomical summer in the northern hemisphere and astronomical winter in the south. Another solstice occurs in December.

MARS is found in Capricornus this month. It reverses direction on 28 June when it reaches a stationary point and goes from direct to retrograde motion. It is still a difficult object for observers far to the north; not only is the ecliptic low in the morning sky, but Mars is somewhat south of the ecliptic. However, from tropical and southern latitudes, the red planet, shining at magnitude –1.3, rises in late evening and gains altitude rapidly.

4 VESTA, the only asteroid that attains naked eye visibility, reaches opposition on 19 June. It can be found shining at magnitude +5.3 in Sagittarius.

JUPITER was at opposition last month and is still visible most of the night. It is best seen from southern and tropical regions but the northern hemisphere is finally getting some good views of the gas giant due to its increased altitude. On 23 June the waxing gibbous Moon passes nearby. Jupiter is by far the brightest object to be found in the faint constellation of Libra.

SATURN is visible most of the night in Sagittarius. It is very near the waning gibbous Moon on the first day of the month. The ringed planet reaches magnitude 0.0 at opposition on 27 June. The following day, at 04:00 UT, it is only 1.8° south of the Full Moon.

URANUS rises after Neptune in the early morning hours. When seen from southern temperate regions, the ecliptic rises steeply in the north east so Uranus, in Aries, quickly gains altitude ahead of the Sun. The northern hemisphere does not get such good views, with Uranus staying relatively low to the horizon at morning twilight. On 10 June, Uranus may be found north of the waning crescent Moon.

NEPTUNE is a morning sky object. It is best seen in tropical and southern temperate latitudes where it rises a little after midnight and is visible in dark skies for several hours before dawn. However, the planet is getting easier to see from the northern hemisphere as it rises ever earlier before the dawn. On 6 June, Neptune is north of the last quarter Moon and on the following day, it reaches west quadrature. The planet arrives at its stationary point on 19 June and changes from direct to retrograde motion. It continues to reside in Aquarius.

A Closer Look at Corona Borealis:
A Golden Crown for a King's Daughter

Brian Jones

The small but distinctive Corona Borealis (the Northern Crown) is located immediately to the east of neighbouring Boötes. Because the constellation lies only a short way north of the celestial equator, the whole of Corona Borealis can be seen from anywhere north of latitude 50°S, rendering it visible to backyard astronomers in South Africa, Australia, New Zealand and most of South America. The two bright stars Arcturus and Izar, both in Boötes, are shown on the chart to help you identify the group.

As you can see from the chart, Corona Borealis takes the form of a pretty semicircle of reasonably bright stars, and belongs to the somewhat elite number of constellations that actually represent the objects or characters they are supposed to depict. Arabic astronomers identified the group as a broken plate (presumably because the stars of the constellation do not form a complete

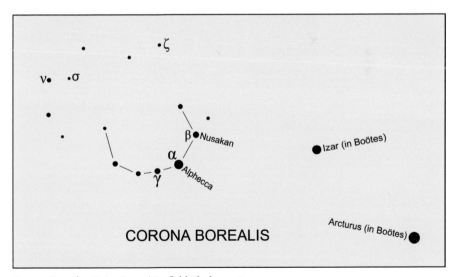

Corona Borealis. (Brian Jones/Garfield Blackmore)

circle). Aboriginal folklore offers us two explanations, one that the group depicts an eagle's nest and the other a heavenly boomerang. However, one of the oldest stories associated with the constellation comes down to us from the Ancient Greeks who identified it as the crown made by the supreme goldsmith Hephaestus at his underwater smithy and which was presented by Bacchus to Ariadne, daughter of King Minos of Crete.

The brightest star in Corona Borealis is magnitude 2.22 Alphecca (α Coronae Borealis), the light from which reaches us from a distance of around 75 light years. Alphecca derives its name from the Arabic for 'break', perhaps echoing the fact that Corona Borealis takes the form of an incomplete circle of stars. The star Alfecca Meridiana in Corona Australis, the southern counterpart of Corona Borealis, derives its name in a similar way.

Somewhat further away from us are magnitude 3.66 Nusakan (β Coronae Borealis), which lies at a distance of 115 light years, and Gamma (γ) Coronae Borealis, the magnitude 3.81 glow of which set off towards us around 150 years ago.

Marking the two 'ends' of the Northern Crown are Theta (θ) and Iota (ι) Coronae Borealis. Theta is a blue giant star, the magnitude 4.14 glow of which reaches us from a distance of around 375 light years. The white giant Iota is a little closer, the light from this magnitude 4.98 star emanating from a distance of a little over 300 light years.

Double Stars in the Northern Crown

Those of you with small telescopes may like to try their hand at resolving the three double stars that together straddle the northern reaches of Corona Borealis. One of these is the binary Sigma (σ) Coronae Borealis, which hovers on the edge of naked-eye visibility a little way to the north east of Iota. Once located, its magnitude 5.70 and 6.70 yellowish components are easily resolved through a small telescope.

Another binary for small telescopes is Zeta (ζ) Coronae Borealis, a system containing two blue-white companion stars of magnitudes 5.00 and 5.95 located at a distance of 470 light years.

Both Sigma and Zeta are comprised of pairs of stars which are actually orbiting each other, a class of object referred to by astronomers as 'binary stars'. These are in contrast to 'optical doubles' in which the two stars happen to lie in the same line of sight as seen from our planet and so only appear to lie close to each other in the sky. The star Nu (v) Coronae Borealis, located just to the east of Sigma, is an optical double, the two stars forming it having orange tints. The brightest is a magnitude 5.20 red giant shining from a distance of around 640 light years, putting it somewhat further away than its slightly-fainter companion, an orange giant star with a magnitude of 5.40, the light from which has taken around 590 years to reach us. The components of Nu are far enough apart to be resolvable in binoculars.

An interesting exercise for binocular observers is to try and count the number of stars visible inside the 'bowl' formed by Corona Borealis. If you can spot a dozen or more then you are doing quite well.

July

New Moon: 13 July
Full Moon: 27 July

MERCURY continues to be visible in the evening sky after sunset but is easier to see from southern latitudes than from the north, where it is already starting to lose altitude in the west. It reaches greatest elongation east on 12 July and is only 2° south of the very young Moon two days later. Aphelion is attained on 20 July and five days later, Mercury reaches a stationary point, moving from direct to retrograde motion. The farther north you are, the earlier in the month Mercury is lost to view but observers in southern temperate regions may be able to see this elusive planet right through to the end of July.

VENUS dominates the evening sky for southern hemisphere observers, blazing away at magnitude −4.2 by the end of the month, and continuing to climb higher every night. Northern latitudes are now at a disadvantage although the bright planet is still prominent in the west. On 9 July, Venus is found only a degree north of the first-magnitude star Regulus in the constellation of Leo, and one week later, produces a pleasing sight with the waxing crescent Moon.

EARTH is at aphelion on 6 July. This marks our planet's farthest distance from the Sun during the year, at 152.1 million kilometres.

MARS is at opposition on 27 July, visible at magnitude −2.8 in Capricornus, but due to the pronounced eccentricity of Mars's orbit, it is closest to the Earth on 31 July. At this time, it is 57.6 million kilometres away and the planet's disk is 24.3 arc-seconds in diameter, making this a very favourable opposition for telescopic observers. For those in southern temperate latitudes, Mars is visible all night. However, for northern hemisphere observers, it only becomes visible around midnight and never gains much altitude.

JUPITER is in Libra and is visible in the evening from both hemispheres, although easier to see from tropical and southern latitudes. On 11 July, it attains a stationary point and reverses direction, moving from retrograde to direct motion. Jupiter has a close encounter with the waxing gibbous Moon on 20 July.

SATURN rises soon after sunset in tropical and southern latitudes and is easy to observe. However, because Sagittarius never gets high in the sky at northern temperate latitudes, Saturn is not easy to see due to its low altitude. The waxing gibbous Moon passes close by the magnitude +0.2 planet on 25 July.

URANUS is best seen from the southern hemisphere. Rising after midnight, it is relatively high in the north before dawn. Observing conditions in northern temperate latitudes are much more difficult because the planet remains rather low to the horizon. On 7 July, Uranus has a close encounter with the waning crescent Moon and on 25 July, the ice giant reaches west quadrature in Aries.

NEPTUNE has two close encounters with the Moon this month. On both 4 July and 31 July, the planet can be found in Aquarius about 3° north of the waning gibbous Moon. Neptune rises before midnight as viewed from the southern hemisphere and is in the northwest by dawn. It still rises after midnight for northern latitudes but is easier to observe in the dark skies preceding sunrise. Because Neptune is never brighter than magnitude +7.8, optical aids are always necessary to see it.

134340 PLUTO reaches opposition on 12 July in Sagittarius. Observing it is challenging as it is at best only magnitude +14.2. Even large amateur telescopes will not resolve a planetary disk but show only a faint spot of light. A detailed star chart is required to differentiate this faint dwarf planet from the myriad of stars inhabiting this region of the sky.

Mars at Opposition

Richard Pearson

Mars comes to opposition in the constellation Capricornus on 27 July 2018 and, at a distance of around 57,745,000 kilometres (35,881,080 miles) will be almost as close to Earth it can be.

Mars is the fourth planet out from the Sun, lying at a mean distance of 227,940,000 kilometres (141,635,350 miles) from our parent star. With a diameter of just 6,780 kilometres (4,212 miles) the planet is little more than half the diameter of the Earth. Mars travels around the Sun once every 687 days which makes the Martian year nearly twice that of our own. Due to its reddish hue when seen in the sky, a number of ancient civilizations associated the planet with bloodshed and war, which led to Mars being named after Ares, the Roman god of War. Because of its distinctive colour, Mars is also known as the red planet.

Since the early years of the space age, many spacecraft have visited Mars in the search for signs of life. The planet has a thin atmosphere comprised mainly of carbon dioxide, the mean atmospheric pressure at the planet's surface being a little over 6 millibars (which varies due to the seasonal changes). This is less than 1% of the Earth's mean sea level pressure. The Martian atmosphere is also significantly thinner than Earth's, so even a slight breeze on the Martian surface can blow the orange surface soil into dusty whirlwinds.

Mainly composed of dry ice (carbon dioxide) with ordinary water ice mixed in with it, the polar ice caps are prominent, and grow and shrink due to the season on Mars (which are twice the length as those on Earth) due to the carbon dioxide ice sublimating (vaporizing) in the Martian summer, thereby revealing the surface, and freezing again in winter.

The axial tilt of Mars is 25 degrees (similar to that of the Earth's 23.5 degrees). Because of its tilt Mars, like the Earth, experiences seasons. The fact that Mars is at a greater distance from the Sun means that the Martian year is around two Earth years long, the consequence of which is that the lengths of the Martian seasons are around twice those of Earth. Martian surface temperatures vary

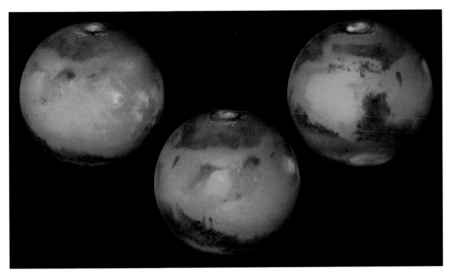

Three images of Mars taken on 10 March 1997 with the Wide Field Planetary Camera-2 on the Hubble Space Telescope (HST). These images were obtained just before the forthcoming opposition of Mars (on 17 March of that year) when the red planet approached to within around 99 million kilometres (61.5 million miles) of the Earth. These pictures were taken during three HST orbits that were separated by about six hours. This sequence of images covers most of the Martian surface, the timing being chosen so that Mars, with its 24-hour 39-minute day, would rotate about 90 degrees between orbits (David Crisp and the WFPC2 Science Team. (JPL/California Institute of Technology), and NASA)

from lows of about $-143\,°C$ at the winter polar caps to highs of up to around $20\,°C$ or more in equatorial summer.

Much of the Martian highlands, made up of craters, mountain ranges and volcanoes, lie in the southern hemisphere of Mars where the crust thickness is around 62 kilometres (39 miles). The crust thins out towards the equator, the northern hemisphere being almost entirely lowlands where the crust thickness is about 38 kilometres (24 miles). It is now believed that the Martian oceans, which existed on Mars a million years ago, filled the Martian lowlands.

Mars has two small irregularly-shaped moons, both discovered by the American astronomer Asaph Hall in August 1877 and subsequently named after the characters Phobos (alarm/panic) and Deimos (dread/terror) who, according to mythology, accompanied their father Ares into battle.

Phobos is the larger and innermost of the two moons, with a mean diameter of 22 kilometres (14 miles) and an orbit which carries it around Mars once every 7 hours and 39 minutes at a height of around 6,000 kilometres (3,730 miles) above the Martian surface. With a mean diameter of just 12 kilometres (7½ miles), Deimos takes a little over 30 hours to orbit Mars at a height of a around 20,000 kilometres (12,430 miles) above the Martian surface.

When Mars is at opposition and at it is closest to us, a telescope of 75 mm (3-inch) aperture will show the main surface markings, although a telescope of 20 cm (8-inch) and above is needed to do useful work. Do not be afraid of using high magnifications (depending on the seeing conditions) but do remember that a small, sharp view of Mars is better than a large, burred one. Remember also that an astronomical telescope will turn the image of Mars upside down so that the South Pole will be uppermost in the field of view.

At or around the opposition date, larger instruments may enable you to see the huge volcano Olympus Mons together with several of the larger Martian craters. Along with the Martian polar ice caps, you should also be able to glimpse dark areas such as Syrtis Major and Acidalia Planitia.

It will also be possible to observe the Martian moons, which will be found to lie just a few arc seconds from the Martian disc. The main difficulty in observing Phobos and Deimos is that their feeble glow can easily be overwhelmed by the glare of Mars itself, Phobos appearing as an 11th magnitude point of light, slightly brighter that 12th magnitude Deimos. Yet with a telescope of at least 100mm (4-inch) aperture and by viewing from a dark observing site with good, clear and moonless skies, you may pull it off. Above anything else, you will need patience and dedication for the search, as did Asaph Hall while he was searching for these visually elusive objects!

Make the most of this month's opposition, since you will have to wait until 15 September 2035 when Mars will be as close (57,147,000 kilometres / 35,509,500 miles) to Earth again.

August

New Moon: 11 August
Full Moon: 26 August

MERCURY is lost to view for the first half of the month as it undergoes inferior conjunction on 9 August but reappears in eastern skies for what is the Mercury's best morning apparition for the northern hemisphere. It changes direction on 18 August, reversing from retrograde to direct motion, and reaches greatest elongation west on 26 August, shining at magnitude –0.5

VENUS is at greatest elongation east on 17 August. It has already started to lose altitude as seen from the northern hemisphere but it continues to climb when viewed from more southerly latitudes. By the end of the month it approaches magnitude –4.5 in the western skies after sunset.

MARS resumes direct motion on 28 August after two months in retrograde. It is becoming easier to observe from the northern hemisphere and is best seen in the dark skies around midnight. Mars remains an easy object from southern latitudes, shining at magnitude –2.5 in Capricornus until 24 August when it moves briefly into the constellation of Sagittarius.

JUPITER attains east quadrature on 6 August in Libra. The interplay of shadows between the planet and its Galilean satellites is at its most pronounced during this time, making for interesting telescopic opportunities. Unfortunately for northern hemisphere observers, the planet is getting nearer to the horizon as astronomical twilight approaches. It sets around midnight in southern temperate latitudes so it is best seen as soon as the sky darkens. The waxing crescent Moon is found nearby on 17 August.

SATURN can be seen in the evening, fading slightly from magnitude +0.2 to +0.3 in Sagittarius. On 21 August, the waxing gibbous Moon passes just north of the planet.

URANUS continues to rise higher in the sky for tropical and southern hemisphere observers for whom it now rises around midnight. For viewers in northern temperate latitudes, the planet, located in Aries, is finally gaining enough height above the horizon for useful observations before morning twilight. On both 3 August and 31 August, Uranus is overtaken by the waning gibbous Moon. Uranus arrives at a stationary point on 7 August and turns to retrograde motion.

NEPTUNE is approaching opposition in Aquarius so it is above the horizon most of the night, rising just before midnight in the northern hemisphere and by mid-evening of southern hemisphere viewers. The nearly-full Moon passes within 2° of Neptune on 27 August.

Life on the Moon

Richard Pearson

With a diameter of 3,475 kilometres (2,160 miles) and orbiting our planet at a mean distance of 384,400 kilometres (238,855 miles) once every 27.3 days, the Moon is the Earth's only permanent natural satellite. Today we know the moon as a lifeless world where the atmosphere is so thin that it provides no protection from either solar radiation or bombardment from meteoroids. The lack of an effective atmosphere means that heat is not held near the surface, so temperatures vary wildly. Consequently, the surface temperature can climb to around 130°C or thereabouts during the day, plummeting to about minus 150°C or less during the lunar night.

Various cultures around the world have long identified animals as well as human faces depicted in the patterns formed from the dark lunar maria and bright cratered highlands, including the romantic and familiar notion of the 'Man in the Moon'. However, it was following the invention of the telescope in 1609 that some observers began to believe that the Moon must have life on its cratered surface.

Image of the Full Moon taken in March 2015. (David Blanchflower)

Although we now know that the lunar maria are extensive flat plains resulting from the spread of volcanic lava during an earlier period of lunar evolution, early observers of the Moon thought that these dark regions of solidified lava were actually seas surrounded by brighter land areas. The German astronomer Johannes Kepler (1571–1630) believed that the telescope had revealed a living world, with extensive oceans and a dense mantle of air.

By the end of the 17th century, the idea of oceans on the Moon had been abandoned, although the notion that life existed there had not. William Herschel, best remembered for the discovery of the planet Uranus in 1781, considered the Moon to be habited by a race of beings he termed as 'Lunarians'. It is on record that he once submitted a paper on lunar mountains to the Royal Society in London and that, before it was accepted, the fifth Astronomer Royal, Nevil Maskelyne, insisted a paragraph relating to 'lunar inhabitants' should be deleted!

Another well-known lunar observer was the German astronomer Johann Hieronymus Schröter who was also sure that the Moon was inhabited. He reported seeing colour changes on the lunar surface and attributed these to areas under cultivation. He also considered some of the formations he observed to be artificial.

Apollo 15 Lunar Module Pilot Astronaut James B. Irwin gives a military a salute while standing beside the deployed U.S. flag in July 1971 during the Apollo 15 lunar surface extravehicular activity (EVA) at the Hadley-Apennine landing site (NASA)

Schröter's ideas seem to have been supported by another German astronomer, Baron Franz von Paula Gruithuisen, who announced in 1822 that he had discovered a real lunar city inside the crater Schröter, which he described as a fortified structure with walls and built by a race of beings he termed the 'Selenites'.

The last advocate of life on the Moon was the famous American astronomer William Henry Pickering who carried out a detailed study of the lunar crater Eratosthenes, situated close to the lunar Apennine Mountains on the south eastern shores of Mare Imbrium. He reported finding a number of strange dark patches around Eratosthenes which displayed regular variations each lunar day. He was convinced that 'vegetation tracts' did exist on the Moon, his ideas being published in The New York Times of 9 October 1921 and in Pickering's final paper on the subject, published in 1924.

Although the Moon is lifeless, American astronauts have landed there during the Apollo missions of the 1960s and 1970s. Neil Armstrong was the first man to set

foot on the Moon's surface in July 1969, describing what he saw as 'magnificent desolation'. The last man to walk on the lunar surface was Eugene 'Gene' Cernan, during the Apollo 17 mission of 1972.

Today, renewed interest in lunar exploration has surfaced again as India, China, and the American space programmes of NASA and SpaceX intend to either land on the Moon, or navigate around the Moon. The first lunar base is sure to follow, and only then will future generations of observers be able to look up at the Moon and describe it as a truly habited world.

September

New Moon: 9 September
Full Moon: 25 September

MERCURY is at perihelion on 2 September and finds itself a degree north of Regulus, the brightest star in the constellation of Leo, three days later. Observers in the northern hemisphere may be able to witness this event but Mercury quickly vanishes into the solar glare for viewers in southern temperate latitudes, reappearing by the end of the month in the evening sky. However, even those in the north lose sight of the planet by mid-month as Mercury undergoes superior conjunction (the third one this year) on 21 September.

VENUS is found just over a degree south of the first-magnitude star Spica in the constellation of Virgo on the first day of the month. Four days later the planet reaches aphelion. Greatest brilliancy, a delicate balance of planetary phase and distance from the Earth, is on 21 September when Venus achieves magnitude –4.6. This bright object is easily visible from southern temperate latitudes all month but it is getting quite low in the north west as seen from regions farther north.

EARTH reaches its second equinox of the year on 23 September. The Sun is directly over the equator on this day, passing from the northern hemisphere to the south hemisphere, and signalling the beginning of astronomical autumn in the north and astronomical spring in southern latitudes. In North America, the traditional name for the Full Moon nearest to the autumnal equinox is 'Harvest Moon'. This year, the Harvest Moon occurs on 25 September.

MARS reaches perihelion on 16 September. Observers on Earth can see the waxing gibbous Moon pass near by Mars four days later. It continues to be well-placed for viewing from tropical and southern latitudes, high in Capricornus at sunset.

JUPITER is getting increasingly difficult to see from northern temperate latitudes. From the southern hemisphere it appears high in the west at sunset but disappears before midnight. The waxing crescent Moon is close by on 14 September. Jupiter is shining at magnitude –1.8 in Libra.

SATURN reaches a stationary point in Sagittarius on 6 September, returning to direct motion. The rings of the planet are at their most open as viewed from the Earth on 18 September and one week later, Saturn reaches east quadrature. This should provide excellent viewing and photographic opportunities, especially for observers in tropical and southern latitudes, as the shadows of the planet, rings and moons are cast at their most extreme angles as seen from our vantage point on Earth.

URANUS is found in Aries. It is usefully placed for observation this month as it approaches its October opposition. The Moon has been making close passes to this planet all year and this month is no exception, with the appulse occurring on 27 September.

NEPTUNE is at opposition on 7 September in Aquarius, shining at its maximum magnitude of +7.8. This is too faint to see with the naked eye, even under dark skies, so binoculars are required. On 23 September, Neptune is 2° north of the waxing gibbous Moon.

A Closer Look at Sculptor:
A Workshop in the Sky

Brian Jones

This group was originally called Apparatus Sculptoris, a name which has since been shortened to Sculptor (the Sculptor). This is one of the constellations devised by the French astronomer Nicolas Louis de Lacaille during the early 1750s (for further details see the article *Nicolas Louis De La Caille: Bringing Order to the Southern Skies* in the *Yearbook of Astronomy 2017*). Taking the form of a curved line of faint stars, none of which are named and none of which are prominent, the fact that this tiny constellation was intended to represent a sculptor's workshop says a great deal for the imagination of the astronomer who devised it.

Sculptor is situated immediately to the east of the neighbouring constellation Piscis Austrinus, the brightest member of which is Fomalhaut, which is included on the chart as a guide. The whole of Sculptor is visible to observers south of

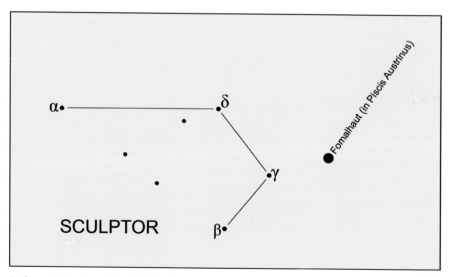

Sculptor. (Brian Jones/Garfield Blackmore)

latitude 50°N, although the absence of any bright stars in the constellation will make it difficult to see by star gazers at mid-northern latitudes unless the skies above their southern horizon are completely dark, clear and free of light pollution.

The brightest star in Sculptor is the blue giant Alpha (α) Sculptoris which shines at magnitude 4.30 from a distance of over 750 light years. Somewhat closer is magnitude 4.38 Beta (β) Sculptoris, the light from which has taken 175 years to reach us. Gamma (γ) Sculptoris is an orange giant, its magnitude 4.41 glow emanating from a distance of 180 light years. The main outline of the constellation is completed by Delta (δ) Sculptoris, the magnitude 4.59 glow of which set off towards us around 140 years ago.

October

New Moon: 9 October
Full Moon: 24 October

MERCURY puts on another fine show at sunset this month for observers in the southern hemisphere. On 5 October, it is 2° north of the first-magnitude star Spica in Virgo. Aphelion is reached on 16 October and then Mercury comes within 3° of the brighter planet Jupiter on 29 October. For the northern hemisphere, this is a poor apparition, with Mercury never gaining much altitude above the western horizon.

VENUS rapidly disappears from northern hemisphere skies but remains visible until mid-month as seen from southern temperate latitudes. Its crescent shape becomes ever thinner until it disappears altogether during inferior conjunction on 26 October. Earlier in the month, on 5 October, Venus changes direction, moving from direct to retrograde motion.

MARS has a very close encounter with the waxing gibbous Moon at 13:00 UT on 18 October, appearing 1.9° south of our natural satellite. It is low in the south east as night falls in the northern hemisphere and sets around midnight but southern observers can see the planet quite high in the sky during the evening hours.

JUPITER is low in the west as darkness falls and is largely lost to view before the end of the month. On 11 October, it is found near the waxing crescent Moon and on 29 October, it has a close encounter with Mercury. Jupiter is still found in the faint constellation of Libra.

SATURN is low in the south west and sets soon after nightfall as seen from northern latitudes. However, it is quite high in the west after sunset in the southern hemisphere and sets around midnight. The waxing crescent Moon passes within 1.8° of first-magnitude Saturn on 15 October at 03:00 UT. Look for the ringed planet in Sagittarius.

URANUS is visible all night in Aries and reaches opposition on 24 October when it shines at magnitude +5.7. However, it will also be about 5° north of the Full Moon on that date, making any naked eye observation almost impossible.

NEPTUNE was at opposition last month so it is still visible most of the night. On 20 October, it is slightly north of the waxing gibbous Moon in Aquarius.

Mission to Mercury: BepiColombo

Richard Pearson

BepiColombo is a joint mission between the European Space Agency (ESA) and the Japanese Aerospace Exploration Agency (JAXA), the aim of which is to orbit and study Mercury with the help of two probes. While the European-built Mercury Planetary Orbiter (MPO) maps the planet, the Japanese Mercury Magnetospheric Orbiter (MMO) will gather information on its magnetosphere.

Among several investigations, BepiColombo will provide us with a complete map of Mercury at different wavelengths as well as charting the planet's mineralogy and elemental composition; determining whether the interior of the planet is molten or not; and investigating the extent and origin of Mercury's magnetic field.

A number of impact craters which are in constant shadow have been found in the planet's polar regions. In spite of the fact that Mercury orbits the Sun at a mean distance of just 57.9 million kilometres (36,000,000 miles), water has been discovered there, and the spacecraft will measure this water concentration in various wavelengths.

The BepiColombo mission is extremely challenging due to Mercury being the closest planet to the Sun and constantly bathed in a stream of energetic charged particles (the solar-wind) constantly streaming out from our parent star. The radiation can be particularly intense following the eruption of solar flares, the strongest of which, Coronal Mass Ejections, are more frequent

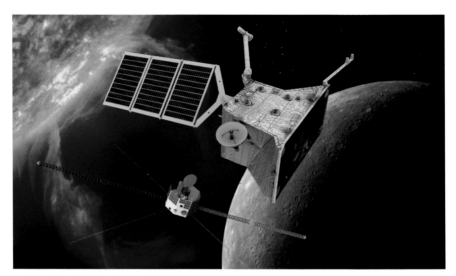

Artist's impression of the BepiColombo mission, showing the Mercury Planetary Orbiter (top) and the Mercury Magnetospheric Orbiter (bottom). (ESA)

during the time of solar maxima when a great many sunspots are on show on the sun's visible surface layer. Fortunately for BepiColombo, the Sun is now going through a quiet phase with sunspot minima expected to occur sometime in 2021.

Another problem are the high temperatures in the Mercury environment and when BepiColombo arrives at the planet in late-2025 it will be subjected to temperatures in excess of 350°C while gathering data.

If all goes to plan, following lift-off from Europe's Spaceport in French Guiana in October 2018 aboard an Ariane 5 launch vehicle BepiColombo will separate from its carrier rocket and enter Earth orbit. Here it will remain for a few days while mission managers carry out instrument checks to ensure that its solar-electric propulsion system is working properly and that the high-gain antenna is communicating correctly.

The spacecraft will then embark on a 7.2-year interplanetary cruise to Mercury and, after a year and a half, will return to Earth where it will perform a gravity-assist manoeuvre and be deflected towards Venus. Two consecutive Venus flybys will reduce its perihelion to approximately Mercury's distance, a subsequent sequence of six Mercury flybys lowering the relative velocity to

1.8 km/s. Four final thruster burns will further reduce the relative velocity to a point in December 2025 when the gravitational pull of Mercury will capture the spacecraft and draw it into a polar orbit without an orbit insertion manoeuvre being required.

Much of our knowledge about Mercury comes from images and science data returned in the 1970s by Mariner 10, and from the robotic spacecraft

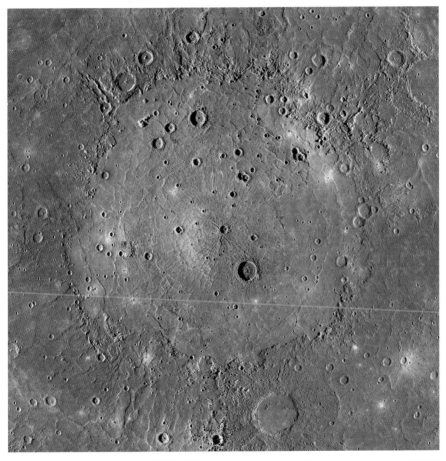

Based on photographs by the MESSENGER orbiter, this Mercury Dual Imaging System (MDIS) mosaic of the mighty Caloris Basin reveals Mercury's youngest large impact basin in its full glory. Caloris is indeed huge - the interior smooth plains have an area of 1.72 million square kilometres (0.66 million square miles), approximately equalling the area of Alaska! (NASA/Johns Hopkins University Applied Physics Laboratory/Carnegie Institution of Washington)

MESSENGER (**Me**rcury **S**urface, **S**pace **En**vironment, **Ge**ochemistry, and **R**anging), the last space probe to visit the planet and which orbited Mercury between 2011 and 2015.

The surface of Mercury is similar in appearance to our Moon in that it is strewn with impact craters and basins. With a diameter of 1,550 kilometres (960 miles) the largest of these, and one of the largest features of its type in the solar system, is Caloris Basin, discovered on images taken by the Mariner 10 probe in 1974. The meteorite impact that created the Caloris Basin occurred during the early formation of the solar system and was so powerful that it caused lava eruptions and left a concentric ring over 2 kilometres (1¼ miles) tall surrounding the impact crater.

Mercury consists of approximately 70% metallic and 30% silicate material, giving it a density second only to that of the Earth. Mercury is much smaller than the Earth and its inner regions are not as compressed so, for the planet to have such a high density its core must be large and rich in iron. What we do not know for certain is whether the core is molten or solid in composition.

Geologists estimate that just over half of Mercury's volume is occupied by its core, surrounding which is a 500-700 kilometre (310-435 mile) thick mantle consisting of silicates. This is topped off with a crust which, based on a mixture of data received from Mariner 10 and Earth-based observation, is estimated to be around 35 kilometres (22 miles) thick. The relatively thin crust is something of a dilemma for planetary scientists. One computer model suggests that, shortly after Mercury formed as a planet around 4.5 billion years ago, another large planet-size body impacted with Mercury, stripping away its crust and leaving behind a substantial iron-rich core and the thin mantle we observe today.

It is hoped that, during its planned encounter with the planet, BepiColombo will provide answers to many of the questions we have about Mercury. In the meantime, backyard astronomers can content themselves simply by catching a fleeting glimpse of Mercury in the twilight sky, either just before sunrise or soon after sunset. This sometimes-elusive little world can be tracked down by checking out and following the information given in the Monthly Sky Notes. However, the small diameter of Mercury means that a moderate size telescope will be needed in order to reveal Mercury's changing phases.

November

New Moon: 7 November
Full Moon: 23 November

MERCURY is at greatest elongation east on 6 November and spends the month losing altitude above the western horizon. On 9 November, the tiny planet is 1.8° north of the first-magnitude star Antares in the constellation of Scorpius, but this event will be difficult to observe from the northern hemisphere. Mercury changes from direct to retrograde motion on 17 November by which time it has vanished from view for northern viewers. It soon disappears for southern observers too as it undergoes inferior conjunction on 27 November. On 29 November, Mercury goes through its fourth and final perihelion of the year.

VENUS passes into the morning sky, visible in the east before sunrise and appearing higher in the sky every morning, brightening all the while. On 14 November, after a little more than a month in retrograde, Venus reaches another stationary point and resumes direct motion. On this day it appears less than half a degree south of first-magnitude Spica in the constellation of Virgo.

MARS moves from Capricornus into Aquarius on 11 November, and on 16 November, it appears 1° north of the waxing gibbous Moon. This is close enough for Mars to be occulted by the Moon as seen from the southernmost regions of Argentina and Chile, and most of Antarctica. Northern hemisphere viewers are finding that Mars has gained some useful altitude in the south after the sky is fully dark but it does set before midnight. Southern observers continue to have the best views, with Mars high in the north at nightfall and setting a little after midnight.

3 JUNO is at opposition on 17 November. It is below naked eye visibility at magnitude +7.5 in the constellation of Eridanus so binoculars or a small telescope will be necessary to see it.

JUPITER is at conjunction on 26 November and is not visible this month. On 20 November, it moves from Libra into Scorpius.

SATURN is low in the west after sunset as it draws closer to the Sun and conjunction next year. On 11 November at around 15:30 UT, it is within 1.5° of the waxing crescent Moon in Sagittarius.

URANUS was at opposition last month and is still aloft most of the night in Aries, not setting until morning twilight. It appears just north of the waxing gibbous Moon on 20 November.

NEPTUNE is an evening sky object. On 17 November, the waxing gibbous Moon passes close by and on 25 November, Neptune arrives at a stationary point, resuming direct motion amongst the background stars of Aquarius.

Edmond Halley: A Question of Pronunciation

Brian Jones

November sees the anniversary of the birth, at Haggerston, near London on 8 November 1656, of the English astronomer, scientist, geophysicist, mathematician and meteorologist Edmond Halley, who is best remembered for computing the orbit, and successfully predicting the return to the inner solar system, of the famous comet that now bears his name.

Although he is remembered primarily as an astronomer (including being the second Astronomer Royal, succeeding John Flamsteed in 1720 and holding the position until his death in 1742), his achievements embrace a wide range of disciplines and include significant discoveries in the fields of mathematics, physics, navigation, meteorology and geophysics. His fame is reflected in the fact that at least two pubs now bear reference to his name along with several streets and roads, an Antarctic research station, a crater on the Moon (and one

Portrait of Edmund Halley c1687 by the Scottish portrait painter Thomas Murray. (Wikimedia Commons)

on Mars) and a mountain here on Earth. However, his legacy may be said to include a particular area of controversy in that his surname is one of the most commonly mispronounced words in astronomy.

During the time that Edmond Halley lived names were often written phonetically – in other words, as they were actually heard. For example, the early registers of Halifax Parish Church in the West Riding of Yorkshire contain a dozen or more different spellings of the now-standardised name Greenwood. These reflect the way the name was pronounced at the time and the way that the parish officials heard, and subsequently recorded, that pronunciation.

Nowadays, of course, the spelling of this and other names have been standardised. However, the problem has not really gone away. For example, ask

Portrait of Samuel
Pepys (1666) by the
English portrait painter
John Hayls. (Wikimedia
Commons)

anyone who lives in Shrewsbury, Leominster or Bicester how often the names of their towns come across verbally and you will realise this!

Anyway, to get back to the issue in hand ... the surname Halley, when applied to the astronomer, is a victim of two particularly common mispronunciations. The first of these, and perhaps the most common, is 'Haley', where the first syllable of the name rhymes with 'hail'. This is an obvious allusion to the rock and roll singer Bill Haley and his backing group The Comets. Clearly this pronunciation is wrong, being a reflection of the common pronunciation of Halley and Halley's Comet in America during the 1950s and 1960s together with the perhaps-obvious (albeit erroneous) association with the astronomer from where the band appear to have derived their name.

The second common mistake occurs when the first syllable of the name is pronounced in the same way as that in the word 'Halloween'. Logically, and taking into account the basic spelling of the name, this pronunciation would appear to be the correct one. However, when we take a closer look at contemporary written versions of Edmond Halley's surname, the *true* pronunciation is clearly revealed to us.

As noted above, prior to the early 19th century, names were often written down and recorded the way they sounded. In other words, and importantly in this context, the way they were pronounced by the people themselves. In the case of Edmond Halley, perhaps the most famous examples of contemporary sources are the diaries of Samuel Pepys (1633–1703). One entry, in which Pepys is praising Halley's work and achievements in navigation, reads:

> "Mr Hawley – May he not be said to have the most, if not to be the first Englishman (and possibly any other) that had so much, or (it might be said) any competent degree (meeting in them) of the science and practice (both) of navigation ..."

This is *clear evidence*, from a *reliable contemporary source*, of the correct pronunciation of the surname of Edmond Halley. From this diary entry, and other phonetic variations of the period, it is clearly evident that the name was pronounced 'Hawley', with the 'Hall' part of the surname rhyming with 'ball'. This is a true and accurate reflection of the pronunciation that Halley himself preferred and which, out of respect to him, the one that we should certainly use today.

December

New Moon: 7 December
Full Moon: 22 December

MERCURY is found in eastern skies before sunrise this month. Best observed from northern temperate latitudes, it is 1.9° south of the waning crescent Moon at 21:00 UT on 5 December. The following day Mercury reaches a stationary point, resuming direct motion. On 15 December, it reaches greatest elongation west. By this time, it is starting to lose altitude above the eastern horizon. Shining at magnitude –0.5, Mercury is 0.8° south of the much brighter Jupiter on 21 December.

VENUS rules the morning sky, reaching greatest brilliancy of magnitude –4.7 on the second day of the month, and climbing ever farther away from the Sun with each passing day. The waning crescent Moon is in close proximity on 3 December and Venus reaches its second perihelion of the year on 26 December.

EARTH arrives at solstice on 21 December. This is the shortest day of the year in the northern hemisphere and the longest day south of the equator, exactly opposite to the June solstice. Astronomical summer begins in the southern hemisphere and astronomical winter arrives in the north.

MARS reaches east quadrature on 3 December, and four days later, it is within half a degree of Neptune. With Mars shining at magnitude +0.1 and Neptune at +7.9, binoculars or a telescope will be necessary to observe this close encounter. Although Mars appears higher in the sky from southern temperate latitudes, viewers in both hemispheres have a good chance to see the red planet this month. On 21 December, Mars moves from Aquarius into Pisces.

JUPITER re-emerges at the end of the month as a morning sky object which is best seen from the southern hemisphere. Northern viewers will find it difficult to catch a glimpse of this planet this month. On 13 December, Jupiter moves from Scorpius to the non-zodiacal constellation of Ophiuchus. Jupiter, shining at magnitude −1.7, has its closest approach to Mercury on 21 December. The following day, it is 5° north of the first-magnitude star Antares.

SATURN is passed by the waxing crescent Moon on 9 December, and is actually occulted as seen from parts of eastern Russia. Saturn is soon lost in the glare of our parent star and is unobservable for most of the month.

URANUS moves from Aries back into Pisces on 3 December. On 18 December, it encounters the waxing gibbous Moon. The planet is well-placed for viewing in the evening sky, not setting until the early morning hours.

NEPTUNE sets around midnight or a little afterwards this month. On 5 December, the planet is at east quadrature and two days later, it is less than half a degree away from the much brighter planet Mars. Neptune is only eighth magnitude so binoculars or a small telescope will be necessary to observe this appulse in the constellation of Aquarius. Neptune's final act for the year is a close encounter with the waxing crescent Moon on 14 December.

Gaia and the Distances of the Stars

Richard Pearson

The European spacecraft Gaia is scheduled to end its five-year mission in or around late-2018 after successfully mapping the positions of more than a billion stars in the Milky Way Galaxy and Local Group. The name of the Gaia mission was originally derived as an acronym from Global Astrometric Interferometer for Astrophysics (GAIA) reflecting the technique of optical interferometry that was originally intended for use on the spacecraft. Although the acronym GAIA no longer applies, the name remains to provide mission continuity.

Gaia mapping the stars of the Milky Way. (ESA/ATG medialab; background: ESO/S. Brunier)

Gaia was launched by Arianespace on 19 December 2013 from Kourou in French Guiana by using a Soyuz ST-B rocket with a Fregat-MT upper stage. The primary goal of the Gaia mission was to obtain data to allow and facilitate further study of the composition, formation and evolution of the Milky Way Galaxy. During its time in space Gaia has performed an all sky survey, mapping the three dimensional position and velocity of all objects down to 20th magnitude and measuring the positions of more than one billion stars with accuracy down to 24 micro-arc-seconds.

Distances to the stars and other galaxies are expressed in a unit of distance known as a 'light year', as measuring the distances in miles would involve numbers so huge that they would be unwieldy. A light year is the distance that a beam of light, travelling at around 300,000 kilometres (186,000 miles) per second, would travel in a year and is equivalent to some 10 trillion kilometres (around 6 trillion miles).

When the first set of positional data was released in 2016, it became clear that many of the published constellation books and star charts we have come

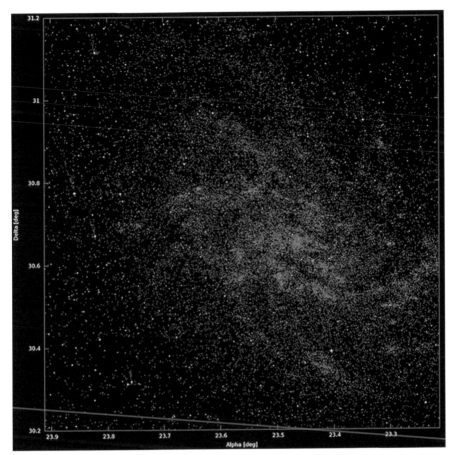

While compiling an unprecedented census of one billion stars in our Galaxy, ESA's Gaia mission is also surveying stars beyond our Milky Way. This image of the galaxy M33 (NGC 598), also known as the Triangulum Spiral, shows tens of thousands of stars detected by Gaia, including a small stellar census in its star-forming region NGC 604. This is a striking example of the mission's potential to detect and characterise stars in nearby galaxies. (ESA/Gaia/DPAC)

to rely on now need to be revised, many stars having been found to be either further away or closer to us than previously measured.

Examples of revisions to stellar distances instigated by the Gaia mission include Polaris (α Ursae Minoris), which has been found to lie slightly further away from us than previously thought, and Spica (α Virginis) which Gaia reveals to lie over ten years closer.

Other examples are the stars that form the celebrated constellation of Orion, the main outline of which comprises the bright stars Betelgeuse (α Orionis), Bellatrix (γ Orionis), Rigel (β Orionis) and Saiph (κ Orionis). The constellation is divided into two halves by a line of three stars that form the Belt of Orion which are, from east to west, Alnitak (ζ Orionis), Alnilam (ε Orionis) and Mintaka (δ Orionis).

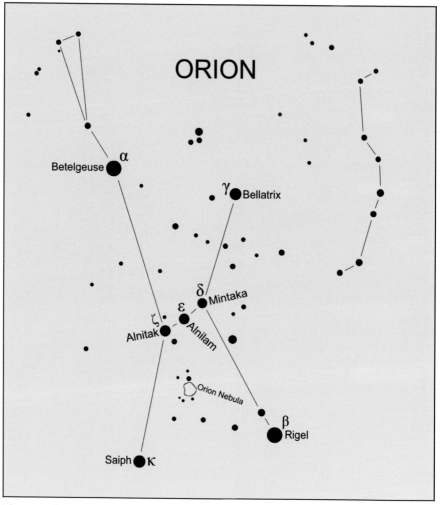

The constellation Orion. (Brian Jones / Garfield Blackmore)

Many of these stars turn out to be further away that previously believed, the best example being Alnilam which is now known to lie more than 500 light years beyond former estimates. Mintaka, on the other hand, is over 200 light years closer to us than earlier measurements suggested.

The Gaia spacecraft has also provided astronomers with a large dataset of measurements to help them in their search for exoplanets – planets that orbit stars other than the Sun. With access to the high-precision results delivered by Gaia, scientists forecast that it will now be possible to detect some tens of thousands of exoplanets out to around 1,600 light-years from the Sun. This will be through the measurement of the effects the planets have on the stars they orbit, specifically the 'wobble' that orbiting planets cause in the motions of their parent stars across the sky.

Comets in 2018

Neil Norman

The year 2018 may well be remembered as the year of the periodic comet because of the 72 comets expected to return to perihelion, no fewer than 57 come within the periodic category, having orbital periods of less than 20 years, and consequently having made many perihelion passages over the years.

This is a problem for the backyard observer who does not possess state-of-the-art equipment because, as a rule, these comets do not become much brighter than 12th magnitude, which means that either a large telescope or CCD equipment is required. They do not become bright targets unless they either have an outburst event (such as Comet Holmes of 2007) or they make close passes to Earth ...

... and luckily for us, later this year will see a close pass by two of these comets.

This being said, it is worth remembering that comets are being discovered on a regular basis and the next Hale-Bopp or Hyakutake could be discovered at any time! You can keep up to date with these discoveries by checking out the British Astronomical Association Comet Section page (**www.ast.cam.ac.uk**) where you will read the prospects for any newly discovered comets.

The Brightest Comets in 2018

COMET	PERIHELION	BEST MAGNITUDE	PERIOD (Years)
185P/Petriew	27 January	11	5.46
C/2016 R2 (PanSTARRS)	9 May	9	20,720
2013 CU129 (PanSTARRS)	24 June	11	4.89
C/2016 M1 (PanSTARRS)	10 August	9	84,299
48P/Johnson	12 August	10	6.54
21P/Giacobini-Zinner	10 September	5	6.62
38P/Stephen-Oterma	10 November	9	38.0
46P/Wirtanen	12 December	3	5.44

The comets listed here should put on a good show, but do remember that comets are very unpredictable and so can either exceed expectations, or fail to live up to expected potential.

Best Prospects for 2018

C/2016 R2 (PanSTARRS) was discovered on 9 September 2016 while still around 6.3 astronomical units (AUs) from the Sun and, during the early part of 2018, should be the brightest comet in the sky. During January this long-period comet will be at its best, hovering at around magnitude 10, and will be well placed for observers in the northern hemisphere. The comet reaches perihelion on 9 May at a distance from the Sun of 2.6 AUs, which is by no means close. The comet remains well placed throughout the rest of the year and remains at around magnitude 10 or 11. The period for this comet is in the order of 20,720 years.

DATE	R.A.	DEC	MAG	CONSTELLATION
1 January 2018	04 23 59	+14 05 44	10	Taurus
1 February 2018	03 59 52	+23 05 08	10	Taurus
1 March 2018	04 02 23	+30 17 36	10	Taurus
1 April 2018	04 29 12	+37 23 09	10	Perseus
1 May 2018	05 15 16	+43 28 34	10	Auriga
1 June 2018	06 22 12	+48 38 59	10	Auriga
1 July 2018	07 44 30	+51 53 28	10	Lynx
1 August 2018	09 31 29	+52 39 20	10	Ursa Major
1 September 2018	10 59 03	+50 42 34	10	Ursa Major
1 October 2018	12 23 45	+47 12 08	10.1	Canes Venatici
1 November 2018	13 37 58	+43 26 58	10.4	Canes Venatici
1 December 2018	14 37 01	+41 00 15	10.6	Boötes

21P/Giacobini-Zinner was discovered in 1900 by the French astronomer Michel Giacobini of the Observatoire de Nice (Nice Observatory) in France. Orbital calculations confirmed it to be a periodic comet and it was believed the following two returns would be unfavourable, and indeed it was not until October 1913 that it was recovered, by Ernst Zinner of the Bamberg Observatory, Germany. The orbit was refined and the period was determined as 6.5 years and not the 6.8 as previously believed.

Image of Comet 21P/Giacobini-Zinner taken from Circolo AStrofili Talmassons (CAST) Observatory on 6 December 2012 by Rolando Ligustri. (*Circle amateur Talmassons, CAST MPC 235, Italy http://cara.uai.it*)

The comet was then linked to the October Draconid meteor shower and, although the comet is at perihelion this year, no significant activity regarding the meteor shower is expected.

Although 21P/Giacobini-Zinner is stable in its orbit, it does show a rapid brightening trend pre-perihelion and fading post-perihelion; in fact, the brightening is reliable to begin some 100 days pre-perihelion and the fading reliably begins some 50 days post-perihelion.

This apparition is very favourable as the comet is at perihelion on 10 September and just a day later makes its closest approach to Earth at 0.39 AU. Visually, the comet should be picked up in June, and by the end of August it may have become a naked eye object. This is expected to continue throughout September although it begins to sink southwards soon after, but experienced observers can hope to follow it until the end of October.

DATE	R.A.	DEC	MAG	CONSTELLATION
15 July 2018	22 10 22	+57 32 47	10	Cepheus
1 August 2018	23 50 51	+65 14 19	9	Cassiopeia
15 August 2018	02 16 53	+65 22 57	8.1	Cassiopeia
1 September 2018	04 54 51	+49 17 52	4	Auriga
15 September 2018	06 04 26	+25 43 45	4	Gemini
1 October 2018	06 49 00	-00 03 01	5	Monoceros
15 October 2018	07 11 16	-16 04 42	6	Canis Major

38P/Stephen-Oterma is making its first return since 1980 this year and reaches its brightest in November as it attains magnitude 9. This object was initially discovered by the French astronomer Jérôme Eugène Coggia on 22 January 1867 and was missed on its following perihelion passage in 1904. It was not until 1942 that it was seen again, this by the Finnish astronomer Liisi Oterma, and identified as the comet first discovered by Coggia in 1867. It was later found to have an orbital period of 38.0 years and was duly recovered again in 1980 on 13 June at magnitude 18. The comet brightened rapidly, attaining a magnitude of 8 in early December, and was followed until April 1981.

With a period between 20 and 200 years, this comet fits the classic Halley-type family of comets. The orbit is also very interesting in that the perihelion point for the comet sits perfectly at the orbit of Mars and its aphelion point very close to the orbit of Uranus which means that close approaches to the aforementioned planets often occur, as was the case in 1903 when the comet passed within 1.6 AU of Jupiter, then again within 1.94 AU of Jupiter in 1982 and within 1.42 AU of Saturn in 1984.

DATE	R.A.	DEC	MAG	CONSTELLATION
1 November 2018	07 11 18	+18 35 26	10	Gemini
15 November 2018	07 41 50	+22 46 54	9.5	Gemini
1 December 2018	08 10 48	+28 36 14	9.5	Cancer
15 December 2018	08 28 22	+34 17 59	9	Lynx

46P/Wirtanen was discovered by the American astronomer Carl Alvar Wirtanen on 17 January 1948. It was the original target for the ESA Rosetta probe although, due to a missed launch window, the new Rosetta mission target became 67P/ Churyumov- Gerasimenko.

Comet 46P/Wirtanen as imaged on 5 February 2008 by Rolando Ligustri from Talmassons, Italy. (*Circle amateur Talmassons, CAST MPC 235, Italy www.castfvg.it http://cara.uai.it*)

46P/Wirtanen is a Jupiter family comet, the consequence of which is that it is usually fairly dim in nature, unless it makes a favourable pass to Earth. One such pass occurs this year as the comet will get down to 11.62 million km (7,220,000 miles) of our planet, this approach being thanks to the comet having approached to within 0.28 AU of Jupiter in April 1972. This pass reduced the comet's perihelion distance from a mean value of 1.6 AU down to 1.26 AU and a period change from 6.7 years down to 5.8 years.

This apparition of 46P/Wirtanen should see it picked up in early September. The comet begins moving rapidly north in November when it should be within binocular range and will remain well placed for northern hemisphere observers in December, when it may attain a magnitude of around 3 and become a naked eye object. The coma will be very large, much as comet 45P/Honda-Mrkos-Padjusakova was in early 2017, and will be best seen in binoculars.

DATE	R.A.	DEC	MAG	CONSTELLATION
3 December 2018	02 26 25	-18 51 48	4.4	Cetus
8 December 2018	02 46 09	-08 29 23	3.8	Eridanus
13 December 2018	03 15 47	+07 23 07	3.4	Cetus
18 December 2018	03 59 39	+27 29 47	3.2	Taurus
23 December 2018	05 00 27	+45 10 53	3.5	Auriga
28 December 2018	06 12 14	+55 37 26	4.1	Auriga
31 December 2018	06 53 57	+58 46 35	4.5	Lynx

Minor Planets in 2018

Neil Norman

We currently know of more than 700,000 minor planets and of these some 450,000 have been allotted a permanent number, due to their orbits being well defined, and of these over 16,000 have been given names ranging from musicians to scientists to mythological characters.

The vast majority of these objects lie within the main asteroid belt, located between the orbits of Mars and Jupiter. However, some of these rocky travellers, known as Potentially Hazardous Asteroids (PHAs), travel around the Sun in larger elliptical orbits that often bring them into close encounters with the planets. To date there are 1,786 known PHAs and around 150 of these are believed to be more than a kilometre in diameter. To qualify as a PHA these objects must have the capability to pass within 8 million km (5 million miles) of Earth and to be larger than 100 metres across.

A large number of smaller asteroids pass close to the Earth on a regular basis. These asteroids can be anything from just a few meters in diameter to several tens of meters wide, and of course, much smaller ones enter the Earths atmosphere on a daily basis and burn up harmlessly as meteors.

The more serious astronomers interested in following these objects should go to the home page of the Minor Planet Center, whose job it is to keep track of these objects and determine orbits for them. This page can be accessed by going to **www.minorplanetcenter.net** where you will see a table of newly discovered minor planets and Near Earth Objects (NEOs). At the top of the page is a search box that you can use to find information on any object that you are interested in, and from this you can obtain ephemeredes of the chosen subject. All-in-all, the Minor Planet Center site is perhaps the one that dedicated asteroid observers should consult on a regular basis.

Four smaller objects that are due to pass within ten lunar distances of Earth between early January and late February are listed here along with their estimated diameters and magnitudes. These are all very faint, so observers will need to use CCD equipment to track them. More accurate ephemeredes can be calculated from your location using the ephemeris section on the MPC home page.

OBJECT	DATE	LUNAR DISTANCE	DIAMETER (metres)	CONSTELLATION	MAXIMUM MAGNITUDE
2015 BW512	15 Jan 2018	5.6	39	Hercules	23
2009 BE	24 Jan 2018	3.2	25	Aries	18.4
2007 DC	2 Feb 2018	8.1	13	Scorpius	25
2016 EP56	27 Feb 2018	9.3	62	Musca	18.8

The first four asteroids to be discovered were Ceres (1801), Pallas (1802), Juno (1804) and Vesta (1807). These objects are all very easy to observe either with binoculars or a small telescope and of these four Ceres, Pallas and Juno are visible from latitudes of around 52°N during 2018, with Ceres attaining 7th magnitude in February.

1 Ceres

With a diameter of 945km (587 miles) Ceres is the largest object in the asteroid belt that lies between Mars and Jupiter. Discovered by the Italian astronomer Giuseppe Piazzi at Palermo, Sicily on 1 January 1801, Ceres was believed to be a planet until the 1850s when it was moved to the classification of dwarf planet.

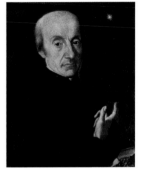

The Italian mathematician and astronomer Giuseppe Piazzi, whose achievements include the discovery of Ceres, the largest of the minor planets and the first to be located. (*Wikipedia*)

DATE	R.A.	DEC	MAGNITUDE	CONSTELLATION
15 Jan 2018	09 26 55	+27 57 27	7.1	Leo
1 Feb 2018	09 12 34	+30 11 02	6.8	Cancer
15 Feb 2018	08 59 22	+31 28 55	7.0	Cancer
1 Mar 2018	08 48 41	+32 03 38	7.3	Cancer
15 Mar 2018	08 43 05	+31 56 44	7.6	Cancer
1 Apr 2018	08 44 19	+31 04 48	7.9	Cancer
15 Apr 2018	08 51 25	+29 56 15	8.2	Cancer
1 May 2018	09 04 54	+28 16 39	8.4	Cancer
15 May 2018	09 20 13	+26 34 15	8.5	Cancer
1 Jun 2018	09 41 54	+24 13 10	8.7	Leo

2 Pallas

Pallas has a diameter of 512 km (318 miles) and was the second asteroid to be discovered when first spotted by the German astronomer Heinrich Wilhelm

Matthias Olbers on 28 March 1802. Pallas's orbit is highly eccentric and its path around the Sun quite steeply inclined to the main plane of the asteroid belt, rendering this object fairly inaccessible to spacecraft.

DATE	R.A.	DEC	MAGNITUDE	CONSTELLATION
1 Dec 2018	12 35 43	-06 39 52	9.1	Virgo
15 Dec 2018	12 58 19	-06 03 01	9.1	Virgo

3 Juno

One of the two largest stony asteroids, Juno was discovered on 1 September 1804 by the German astronomer Karl Ludwig Harding.

DATE	R.A.	DEC	MAGNITUDE	CONSTELLATION
1 Aug 2018	02 48 05	+10 00 52	9.4	Cetus
15 Aug 2018	03 10 37	+09 53 03	9.2	Cetus
1 Sep 2018	03 34 43	+08 55 22	8.8	Taurus
15 Sep 2018	03 50 38	+07 25 33	8.6	Taurus
1 Oct 2018	04 02 47	+04 58 00	8.2	Taurus
15 Oct 2018	04 06 51	+02 18 49	7.9	Taurus
1 Nov 2018	04 03 03	-01 01 59	7.6	Eridanus
15 Nov 2018	03 54 13	-03 16 29	7.4	Eridanus
1 Dec 2018	03 42 17	-04 33 30	7.6	Eridanus
15 Dec 2018	03 34 33	-04 20 02	7.8	Eridanus

6 Hebe

Hebe was discovered by the German amateur astronomer Karl Ludwig Hencke on 1 July 1847 and is most likely to be the parent body of the H type ordinary chondrites, the most common type of meteorite and which account for about 40% of all meteorites hitting Earth.

DATE	R.A.	DEC	MAGNITUDE	CONSTELLATION
1 Sep 2018	05 39 51	+08 51 21	9.9	Orion
15 Sep 2018	06 03 05	+08 04 15	9.8	Orion
1 Oct 2018	06 25 28	+06 53 07	9.6	Monoceros
15 Oct 2018	06 40 27	+05 43 15	9.5	Monoceros
1 Nov 2018	06 51 26	+04 22 57	9.2	Monoceros
15 Nov 2018	06 53 28	+03 34 56	9.0	Monoceros
1 Dec 2018	06 47 28	+03 20 08	8.7	Monoceros
15 Dec 2018	06 35 55	+03 56 23	8.5	Monoceros

7 Iris

The first of ten asteroids discovered by the English astronomer John Russell Hind, Iris was first spotted on 13 August 1847. Iris is of a stony composition and ranks as the fourth brightest object in the asteroid belt.

The ten asteroids discovered by English astronomer John Russell Hind include Iris and Flora, both of which came to light in 1847. (*Wellcome Library, London. Photograph by Maull & Polyblank*)

DATE	R.A.	DEC	MAGNITUDE	CONSTELLATION
15 Jan 2018	02 25 48	+15 47 28	8.8	Aries
1 Feb 2018	02 52 27	+16 48 35	9.2	Aries
15 Feb 2018	03 17 39	+17 51 22	9.4	Aries
1 Mar 2018	03 45 03	+18 55 08	9.6	Taurus
15 Mar 2018	04 14 03	+19 52 00	9.8	Taurus
1 Apr 2018	04 50 49	+20 43 43	10.0	Taurus
15 Apr 2018	05 21 50	+21 06 07	10.1	Taurus

8 Flora

John Russell Hind discovered Flora on 18 October 1847. Named after the Latin goddess of flowers and gardens, and with a mean diameter of 128 km (80 miles), Flora is the innermost large asteroid to orbit the Sun.

DATE	R.A.	DEC	MAGNITUDE	CONSTELLATION
15 Jan 2018	06 39 04	+22 14 59	8.6	Gemini
1 Feb 2018	06 26 27	+23 32 03	9.2	Gemini
15 Feb 2018	06 24 16	+24 17 44	9.6	Gemini
1 Mar 2018	06 29 29	+24 48 34	9.9	Gemini
15 Mar 2018	06 40 51	+25 04 55	10.3	Gemini

101955 Bennu

One last object, 101955 Bennu, is well placed and visible from January to June and is the subject of the NASA OSIRIS-REx mission. This is due to encounter the asteroid in August of this year following which it will return samples back to Earth in 2023.

Bennu was discovered by the Lincoln Near-Earth Asteroid Research (LINEAR) Project on 11 September 1999 and is a roughly-spheroidal carbonaceous asteroid around 500 metres in diameter belonging to the Apollo group of asteroids. It also is a potential future Earth impactor and is listed on the Sentry Risk Table with the second-highest rating on the Palermo Technical Impact Hazard Scale. Because there is a 1-in-2,700 chance of a collision with Earth at some point during the late 22nd century, a mission to explore this object is crucial for future generations. If an impact by an asteroid of this size were to occur, the expected kinetic energy associated with the collision would be equivalent to around 1,200 megatons of TNT.

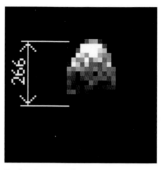

Radar image of 101955 Bennu. (*Arecibo Observatory and JPL*)

DATE	R.A.	DEC	MAGNITUDE	CONSTELLATION
2 Jan 2018	13 56 35	-05 43 46	22.5	Virgo
7 Jan 2018	14 09 36	-07 28 09	22.5	Virgo
12 Jan 2018	14 22 21	-09 09 01	22.5	Libra
17 Jan 2018	14 34 51	-10 46 15	22.5	Libra
22 Jan 2018	14 47 04	-12 19 52	22.5	Libra

Meteor Showers in 2018

Neil Norman

On any given night one can expect to see several meteors dash across the sky. Quite often these will be 'sporadic' meteors, meaning that they can literally appear at any time and from any direction. These are usually just pieces of space debris, ranging from grain-sized pieces of rock to materials lost during rocket launches or space walks. However, at certain times of the year the Earth encounters more organised streams of debris that produce meteors over a regular time span and which seem to emerge from the same point in the sky. These are called meteor showers.

These streams of debris are the remnants of comets that have made repeated passes through the inner solar system. The ascending and descending nodes of their orbits lie at or near the plane of the Earth's orbit around the Sun, the result of which is that at certain times of the year the Earth encounters and passes through a number of these swarms of particles. The term 'shower' must not be taken too literally, as even the strongest annual showers will generally only produce one or two meteors a minute at best, this again depending on the lunar phase and on what time of the evening or morning that you are observing.

A table of the principal meteor showers for 2018 is given here, the information given including the name of the shower; the period over which the shower is active; the ZHR (Zenith Hourly Rate); the parent object from which the meteors originate; the date of peak activity; and the constellation in which the radiant of the shower is situated. Most of this is self-explanatory, although the ZHR may need a little explanation. The Zenith Hourly Rate is the number of meteors you may expect to see if the radiant (the point in the sky from where the meteors appear to emerge) is at the zenith (or overhead point) and if the observing conditions were perfect and included dark, clear and moonless skies with no form of light pollution whatsoever. However, the ZHR should not to be taken as gospel, and you should not expect to actually see the quantities stated, although 'outbursts' can occur with significant activity being seen.

Noting down the colours of meteors will tell you something about their composition. For example red is nitrogen/oxygen, yellow is iron, orange is sodium, purple is calcium and turquoise is magnesium. To avoid confusion with sporadic meteors, be sure to have a star map with you, so you can compare the trajectories against it and, hopefully, trace their origins back to the shower radiant.

Meteor Showers in 2018

SHOWER	DATE	ZHR	PARENT	PEAK	CONSTELLATION
Quadrantids	1 Jan to 5 Jan	120	2003 EH1 (asteroid)	3/4 Jan	Boötes
Lyrids	16 Apr to 25 Apr	18	C/1861 G1 Thatcher	22/23 Apr	Lyra
Eta Aquarids	19 Apr to 28 May	30	1P/Halley	6/7 May	Aquarius
Perseids	17 Jul to 24 Aug	80	109P/Swift-Tuttle	12/13 Aug	Perseus
Orionids	16 Oct to 27 Oct	25	1P/Halley	21/22 Oct	Orion
Taurids	20 Oct to 10 Dec	5	2P/Encke	5 Nov	Taurus
Leonids	15 Nov to 20 Nov	Varies	55P/Tempel/Tuttle	17/18 Nov	Leo
Geminids	7 Dec to 17 Dec	75+	3200 Phaethon	13/14 Dec	Gemini

Quadrantids

The Quadrantids is a significant shower which rivals the July/August Perseids, and at its peak observed rates can reach 50 meteors an hour. The radiant lies east of the star Alkaid in Ursa Major and the meteors are rapid, reaching speeds of 40 km/s or more, although a drawback to this shower is that the period of maximum occurs over a period of just 2 or 3 hours. The parent object has been tied down to the Amor asteroid 2003 EH which is likely to be an extinct comet. Unfortunately an almost-full moon will drown out all but the brightest of the Quadrantid meteors this year.

Lyrids

These meteors move quite rapidly at speeds of up to 48 km/s and the rates vary, but typical values of 10 to 15 per hour are recorded. The maximum falls on the night of 22 April with the radiant lying near the 0.03 magnitude star Vega in the constellation Lyra. The parent of the shower is the long-period comet C/1861 G1 Thatcher, which last came to perihelion at 0.92 AUs on 3 June 1861 having passed Earth at a distance of almost 50 million kilometres (31 million miles) in May of that year. The period of the comet is 415 years and it will next approach perihelion in 2280. The moon will be at around first quarter and will set shortly after midnight, leaving dark and hopefully cloud-free skies for what could be a good display.

Eta Aquarids

One of the two showers associated with 1P/Halley and which is active for a full month between 19 April and 21 May. The radiant lies just to the east of the star Sadalmelik in Aquarius and maximum activity will occur pre-dawn on 7 May with up to 30 meteors per hour expected. These will be travelling at 65 km/s, the high speed due to the parent comet being in a retrograde orbit resulting in the debris entering the atmosphere head-on. The waning Moon around time of peak activity will result in only the brighter meteors being readily visible this year.

Perseids

Perseid meteor captured by Guy Wells on 12 August 2014. (Guy Wells)

A gorgeous sight to see! I have personally noted a variation in yearly amounts. For example, in the years 2012 and 2013 the rates were as one would realistically expect, although 2016 was somewhat lack-lustre – even taking into account a 69% illuminated lunar disc the rates before this were notably down.

These are also fast moving meteors, clocking in at 58 km/s due again to the parent comet (109P/Swift-Tuttle) being in a retrograde orbit. Up to 80 meteors per hour can be seen at their peak, which occurs on 12/13 August, and large fireballs are often observed, with some even seen to cast shadows. A note of caution though – there are a few other showers active at around this time, so inexperienced observers should ensure they follow the trajectories of any meteors seen back to the radiant point in the northern reaches of Perseus. Early evening sees the setting of the thin crescent moon leaving dark skies for what may be an excellent show.

Orionids

The Orionid meteor shower is the second annual shower that is associated with 1P/Halley and occurs between 16 and 27 October with the peak night occurring on 21 October. Of the two showers, this one is the strongest with typical ZHRs of somewhere around 20 to 25 meteors an hour. The velocity of the meteors entering the atmosphere is a speedy 67 km/s.

The radiant of this shower is situated a little way to the north of the red super giant star Betelgeuse in the shoulder of Orion the Hunter, and the Orionids are best viewed in the early hours leading up until dawn when the constellation is at its highest. This year sees an unfavourable display, the nearly full Moon blocking some of the fainter meteors, although the brighter Orionids may still produce a good show.

Taurids

This shower is associated with periodic comet 2P/Encke (last at perihelion in early-2017) and which is the comet with the shortest known orbital period (3.3 years). The stream of debris left by the comet is truly vast, and is by far the largest in the inner Solar System, and with particles of a larger mass than the other showers, fireballs are widely reported with this shower. The reason for this is believed to be that a much larger comet fragmented some 20,000 to 30,000 years ago, leaving behind a large piece that we know today as 2P/Encke, together with a huge amount of smaller debris. In fact, the Tunguska event of 1908 may very well be related to this stream, though this remains a heated debate even today, some 110 years after.

The radiant for this shower lies a little to the south of the Pleiades open star cluster. The ZHR is low, and only five or so can be realistically expected, although they are worth waiting for as they appear to glide across the sky at a sedate pace of around 27 km/s (or in old money, around 60,000 miles per hour). At the time of maximum activity the thin crescent moon will have set during the early part of the evening leaving behind dark skies for viewing.

Leonids

This is a fast moving shower with an atmosphere impact speed of 72 km/s and which also contains a lot of larger sized pieces of debris, with diameters in the order of 10mm and masses of around half a gram. These can create lovely bright meteors that may attain magnitude –1.5 or better. It is interesting to note that each year around 15 tonnes of material is deposited over the planet from the stream.

The parent of the Leonid shower is the periodic comet 55P/Tempel Tuttle which orbits the Sun every 33 years and which was last at perihelion in 1998 and is due to return in late-May 2031. The Leonid radiant is located a few degrees to the north of the bright star Regulus in Leo. The Zenith Hourly Rates vary due to the Earth encountering material from different perihelion passages of the parent comet. For example, the storm of 1833 was due to the 1800 passage, the 1733 passage was responsible for the 1866 storm and the 1966 storm resulted from the 1899 passage. The 2001 and 2002 showers were from the 1767 and 1866 passages respectively. This year, the waxing gibbous moon will set shortly after midnight leaving fairly dark skies for what could be a good early morning show.

Geminids

This shower was first recorded in 1862 which indicates that the shower debris has been perturbed into its current orbit by Jupiter. The parent of the shower is the object 3200 Phaethon, an asteroid that is in many ways behaving like a comet. Discovered in October 1983, this rocky 5-kilometre wide object is classed as an Apollo asteroid and has an unusual orbit that takes it closer to the Sun than any other named asteroid. Classified as a potentially hazardous asteroid (PHA), 3200 Phaethon made a relative close Earth approach of 0.069 AU (10.3 million kilometres / 6.4 million miles) on 10 December 2017.

Geminid meteors travel at speeds of 35 km/s and disintegrate at heights of around 40 kilometres above the Earth's surface. This shower radiates from a point close to the bright star Castor in Gemini and is considered by many to be the best of the year. An interesting note on the shower is that the numbers seem to be increasing each year, reports in 2014 giving a ZHR of 253 regardless of Moon interference! As far as the 2018 shower is concerned, the first quarter moon will set shortly after midnight leaving dark skies for what should be an excellent early morning display.

Article Section

Astronomy in 2017

Rod Hine

ALMA Discoveries – Oldest and Coldest

The Atacama Large Millimeter Array (ALMA), situated high in the wilds of Chile, has produced much valuable data since it became fully operational in 2013 and has become one of the busiest and over-subscribed telescope systems in the world. Two of the most notable recent achievements have been the observations of the coldest known spot in the universe and the discovery of the oldest dust. The huge array, eventually to comprise 66 dishes of 12 metre and 7 metre diameter, achieves unprecedented sensitivity and resolution at wavelengths from 0.3 mm to 9.6 mm. Located 5000 metres above sea-level in one of the driest places on earth it will surely provide even more fascinating results in the future.

The prevailing temperature across most of the sky appears to be 2.7 K which, just above absolute zero, is due to the all-pervasive cosmic microwave background radiation (CMB), discovered in 1965 by the American radio astronomers Arno Penzias and Robert Woodrow Wilson. This radiation represents the afterglow of the Big Bang and fills the whole of space.

Imagine the surprise in 1995 when astronomers found a small region in the Boomerang Nebula, situated around 5000 light years away in Centaurus, which appeared to have a temperature of only 0.1 K. In that region, something must be absorbing the CMB and so must be much colder still. Wouter Vlemmings from Chalmers University in Sweden and Raghvendra Sahai from JPL, Pasadena, used ALMA to investigate the Boomerang Nebula. They discovered a huge spherical shell of gas expanding at immense speed from the central star in the heart of the nebula. The total quantity of gas is about 3.3 solar masses, and within that shell are two smaller lobes of gas, also expanding at great speed. It is the rapid expansion of this gas that has cooled it and allowed it to absorb heat from the CMB. The most likely cause is the demise of a binary star system where the larger star died first and began to swell, followed by a violent merger which expelled the smaller lobes. It is estimated that this happened around

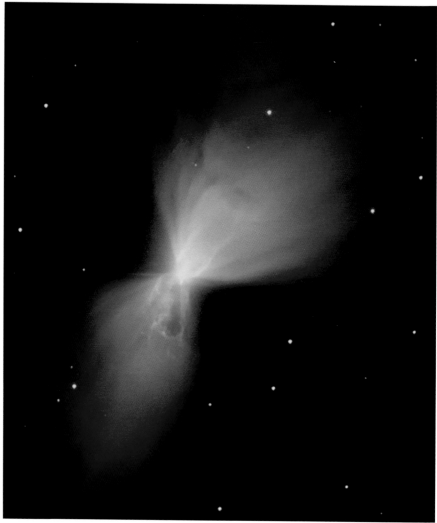

Image of Boomerang Nebula taken by Hubble Space Telescope. (NASA)

1,000 years ago and, of course, the gas will eventually warm up and assume the same temperature as the CMB. It is just a matter of luck that this was seen at all.

The amazing resolution of ALMA was again put to good effect by Nicolas Laporte of University College, London while looking at a very early star-forming galaxy dating back to a time when the universe was only 600 million years old. Amongst those early stars, they found huge amounts of dust, many millions of solar masses in fact. Assuming that dust is largely forged in supernova explosions this could help the understanding of the way the early universe looked. Dust is an essential building block for the creation of complex molecules, planets and even ourselves. Perhaps further observations with ALMA will enable us to see back even further in time to those stars that formed entirely from hydrogen and helium, to galaxies in pristine space with no contamination from supernovae.

New Phenomena

However much we think we know nature can still surprise us with an unexpected observation or a puzzling discovery. In May 2017, NASA reported that scientists monitoring the Cassini spacecraft were mystified by eerie whistling sounds beamed back to Earth during the spacecraft's risky passage between the Saturn's rings and the surface of the planet. Based on previous observations, they had expected a lot of popping and crackling sounds due to dust particles. Instead, the moments of passing through the ring plane were strangely quiet apart from a distinct whistling noise. NASA went so far as to post an audio file and spectrograms of the event. Cassini was due to make a number of passes between the rings and the planet's surface in the following months before the scheduled end of the mission in September 2017. It is suspected that the whistling noise may be some kind of plasma wave effect, masked on previous recordings by the sounds due to dust particles. Exactly what it is, and why there were so few dust particles in this region is yet to be explained.

In April 2017, a group led by Mansi Kasliwal of Caltech published a paper which revealed a totally new class of event seen by the Spitzer Space Telescope. The team are using the short-wave infrared camera (IRAC) to investigate variable and transient IR sources in galaxies. During the early stages of a five-year study, a search of 190 nearby galaxies identified many variable infra-red sources including 43 transient events. Further investigation showed that of these, 21 were known supernovae, 4 were in the luminosity range of novae, and 4 had optical counterparts, leaving 14 whose intensity was between that of supernova and nova and unrelated to any optical observation. Even in the infrared, the light-curves and wavelengths do not correspond to any other phenomena known at present. The new objects have been dubbed SPRITEs and their name must represent a high point in acronym creation (eSPecially Red Intermediate-luminosity Transient Events).

The most intensively studied of these new objects, called 14ajc, located in the spiral galaxy M83, appears to be within a cloud of warm molecular hydrogen that glows strongly in the infrared. Various explanations have been considered and outlined in the paper. Possibly a close interaction or collision of two newly formed stars caused a shockwave that heated the gas cloud, or alternatively it may be a supernova that was completely hidden within a dense dust cloud, or a kind of 'failed supernova' that collapsed into a black hole before the usual shock-wave 'bounce' could tear the star apart. The rate at which luminosity decays is very variable so there may be several different types of interactions causing the 14 SPRITEs identified so far.

Whatever the origin of these events, there is enough interest to justify further work. Sadly, the Spitzer Space Telescope can no longer operate at

Artist impression of the Spitzer Space Telescope in front of a brilliant, infrared view of the plane of the Milky Way Galaxy. (NASA/JPL)

maximum sensitivity in longer wavelengths since its cryogenic liquid helium ran out in 2009, so it is unlikely that Spitzer can do more than just identify potential SPRITEs as this observation programme continues. However, it is hoped that the James Webb Space Telescope, due to launch in October 2018 with its promised excellent infrared performance, could open a whole new range of possibilities in the study of SPRITEs.

Exosystems

There can be few areas of astronomy that are making such steady progress as the study of exoplanets. With the current total standing over 3,600 and even extended to include firm evidence of 'exomoons', we are now beginning to develop far-reaching but realistic models of distant systems. Some are quite alien, others may bear some resemblance to our Solar system, but all of them help us to understand our place in the cosmos.

Most exoplanets have been found by continuous monitoring of the light from stars and looking for small dips in the light level as the planets transit across the face of the star. Although this method can only detect the small proportion of stars that have a transit visible from the Earth, it is possible to monitor many stars simultaneously. The Kepler Space Observatory works in this way and accounts for about a third of confirmed exoplanets so far found. Once a periodic transit has been identified it is then possible to examine changes in the spectrum during the transit and thus get some idea of the composition of any atmosphere of the planet. The Hubble and Spitzer telescopes have been very successful in this aspect of exoplanet research as they can perform precision spectroscopy in a wide range of wavelengths.

Exoplanets that do not have a visible transit can be found by looking for the slight wobble of the parent star causing a change in the spectral lines due to the Doppler effect. This is more difficult than the transit method but has the potential to find more candidates, especially larger planets close to their stars in systems closer to Earth. Several land-based telescopes are engaged in this work such as the High Accuracy Radial velocity Planet Searcher (HARPS) at La Silla, Chile and the High Resolution Echelle Spectrometer (HIRES) instrument used on the Keck Telescopes.

In March 2017 the team led by Michael Gillon of the University of Liege, Belgium, announced the discovery of no fewer than seven planets in orbit

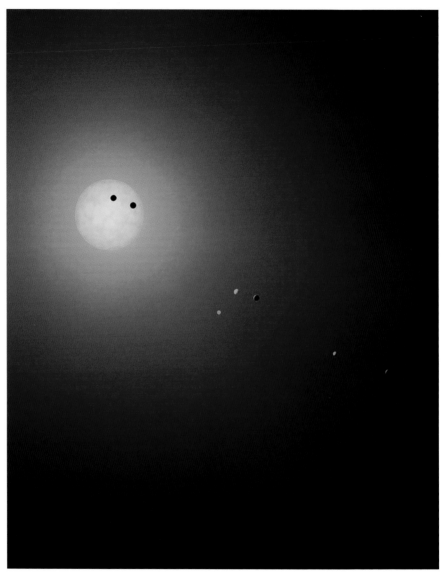

This illustration depicts the tiny system of seven planets orbiting TRAPPIST-1 as they might appear if viewed from Earth using an incredibly powerful, albeit fictional, telescope. The image shows the sizes and relative positions of the planets, each of which is close to the size of Earth, correctly to scale. Their orbits all fall well within what, in our own Solar System, would be the orbital distance from the Sun of our innermost planet, Mercury. (NASA, JPL-Caltech, Spitzer Space Telescope, Robert Hurt)

around a small red star named TRAPPIST-1, some 40 light years distant from Earth and not much bigger in size than our planet Jupiter. Exoplanets are discovered on almost a daily basis but what was remarkable about TRAPPIST-1 was the fact that three of the seven planets were within a possible habitable zone. Although all the planets are very close to the star, even closer than Mercury is to the Sun, the star is small and dimmer than the Sun. Given that such small stars may be prone to X-ray and UV flares it would require very special conditions to allow life to flourish, although it has been suggested that even a relatively thin atmosphere might provide protection for primitive life to survive. The next step is to use the Hubble Space Telescope to look for any signs of atmospheres around these planets. Hubble, and its successor the James Webb Space Telescope, should be able to identify the signature spectrum of the planets' atmospheres, if any, and form a view on the likelihood of life thereon.

Dark Matter, Dark Energy … and Dark Thoughts ..?

By way of contrast to the exosystems industry, the quest for dark matter seems to have stalled and the nature of dark energy remains a mystery. There is little doubt that the Big Bang standard model is well grounded in both theory and observational evidence but, tantalizingly, it still relies on not one, but two components whose physical existence has not yet been discovered. The prevailing view is that the universe, space and time all burst into existence in an instant of unimaginable temperature and density known as the Big Bang, and that the universe we see today is the result of the subsequent expansion and cooling of that primordial fireball. With our knowledge of nuclear physics, it is possible to explain the formation of the hydrogen and helium in the proportions that make up the vast majority of the matter in the universe. Subsequent cooling and formation of stars and galaxies explains the synthesis of the heavier elements, some within supernovae. Also predicted is the presence of the remarkably uniform cosmic background radiation (CMB) which arose from a time around 377,000 years after the Big Bang. It seemed that all could be explained by Einstein's General Relativity, even the expansion of the universe discovered by Hubble in the 1920s.

However, we know now that things are not nearly so simple. In 1933, Fritz Zwicky found that the motion of galaxies within the Coma Galaxy Cluster indicated the presence of unseen matter. Observations of the speed of stars

This false-colour mosaic of the central region of the Coma Galaxy Cluster combines infrared and visible-light images to reveal thousands of faint objects. Appearing here as faint green smudges, subsequent observation has revealed them to be dwarf galaxies belonging to the cluster. Two large elliptical galaxies, NGC 4889 and NGC 4874, dominate the central regions of the cluster. The mosaic combines visible-light data from the Sloan Digital Sky Survey (colour coded blue) with long- and short-wavelength infrared views (red and green, respectively) from NASA's Spitzer Space Telescope. (NASA/JPL-Caltech/L. Jenkins-GSFC)

in the disks of galaxies by the American astronomers Vera Rubin and W. Kent Ford, in the 1960s and 1970s, showed that something was wrong. The stars towards the edge are moving much too fast to stay bounded by the gravity of the visible stars. Either there is a substantial amount of missing mass to hold the galaxies together, or there is something wrong with the laws of gravity. The latter was, and perhaps still is, unthinkable for most physicists, so the concept of 'dark matter' was used to explain the results. All that remained was to deduce the properties of such dark matter and devise experimental means to observe it.

In the 1990s, evidence began to accumulate to show that the expansion of the universe was not constant, as had been expected, but was actually accelerating. In 1998, Michael Turner of the University of Chicago coined the term 'dark energy' as an echo of Fritz Zwicky's 'dark matter'. Thus was set the scene for the 'Lambda-CDM' model. The Lambda refers to the parameter Λ in Einstein's formulation of General Relativity which represents dark energy, and CDM stands for 'Cold Dark Matter'. Such is the power of this model that observations of the CMB by the Planck spacecraft allows the calculation of the relative constituents of the universe, namely 68.3% dark energy, 26.8% dark matter and 4.9% ordinary (baryonic) matter. This is clearly not an ideal state of affairs where the nature of more than 95% of the mass and energy in the universe is of unknown origin, however well all the rest of the theory works.

Recently a team at the Max Planck Institute led by Natascha M. Förster Schreiber used the Very Large Telescope (VLT) in Chile to make exceptionally detailed measurements of six massive galaxies formed about 10 billion years ago. In a letter published in Nature in early 2017 they showed that, in these very early galaxies, the rotation of the outer stars decreases with radius. The conclusion is that such galaxies are dominated by baryonic matter with little contribution from dark matter. Could this mean that dark matter behaved differently in the high-redshift distant regions of the universe? Is it possible that dark matter did not exist then, but does now? To date, numerous experiments have failed to show any kind of direct evidence of the nature of dark matter, and dark energy has proved equally elusive, so it is not surprising that some cosmologists have begun to think the unthinkable after all, no matter how controversial such ideas may once have been.

One of the first suggestions to account for dark matter was the Modified Newtonian Dynamics (MOND) proposed in 1983 by the physicist Mordehai Milgrom, now at the Weizmann Institute in Israel. He suggested that at very small acceleration values, the gravitational force varies inversely with the radius, rather than with the inverse square of radius as in classical Newtonian gravity. While this can work well to account for the behaviour of stars in galaxies, it does not account for the way galaxies move in clusters and it does not seem to lead to any satisfactory cosmological model. Subsequent forms of modified gravity proposed by Jacob Bekenstein in 2004 and John Moffat from 2006 to 2009 have successfully woven relativity into the theories but have not achieved recognition.

More recent alternative theories of gravity are the Emergent Gravity of Erik Verlinde and the Causal Set Theory of Helen Fay Dowker, both of which take quantum considerations into account at the outset. Verlinde, of the University of Amsterdam, believes that spacetime and gravity emerge as a natural consequence of the quantum entanglement at the microscopic level. Put very crudely, the more stuff there is, the more quantum entanglement binds everything together and the greater the resulting 'gravity'. Interestingly, Verlinde's reasoning automatically gives rise to the extra 'dark' gravitational forces, thus obviating the need for dark matter as such. A team led by astrophysicist Margot Brouwer of Leiden University has put it to the test by measuring the distribution of gravity around some 33,000 galaxies and found that the results agree well with his theory. The important advance over the conventional dark matter explanation is that in that case, the parameter has to be adjusted to the observation. Verlinde's theory predicts the results directly without the need for 'free parameters'. Whether the theory will stand up to further tests under more general conditions is yet to be seen, but this must surely be a significant contribution along the path to an understanding of quantum gravity.

In the case of dark energy, there is a sense that now, at the time of writing in 2017, more attention is being given to alternative ways to formulate cosmology theories from completely new perspectives. One approach starts from thermodynamics and may be a clue to understanding the nature of the cosmological constant and the accelerating expansion of the universe towards an empty and featureless void, the so-called de Sitter space which seems to be the eventual fate of the universe. In a paper published in April 2017, Sean Carroll and Aidan Chatwin-Davies of the California Institute of Technology in Pasadena consider how the entropy of the expanding universe changes with time. Entropy can be regarded as a measure of orderliness of a system. The atoms and molecules within a solid body are rigidly fixed and therefore have low entropy whereas the randomly moving molecules in a hot gas have higher entropy. The second law of thermodynamics says that, within a closed system, entropy will increase until it reaches a maximum when all movement is entirely random and there is no order.

Carroll and Chatwin-Davies draw parallels with the state of the universe considered as a closed system. Initially the entropy will be low due to the existence of ordered structures such as stars and galaxies. Because the universe contains a fixed amount of energy and cannot exchange energy with any other system the entropy will inevitably increase until all matter is dispersed and space becomes a flat, featureless and empty void. It is this inexorable increase of entropy that drives the ever-accelerating rate of expansion. In effect, the authors can show that what we call dark energy arises as a consequence of applying the principles of thermodynamics without the need to use Einstein's field equations or even make any assumptions about the sign or value of the cosmological constant. Not only that, they can derive sensible values for Hubble's Constant and the value of Λ. Previous attempts to find Λ from Einstein's field equations tended to give a value some 120 orders of magnitude too large!

Solar System Exploration in 2017

Peter Rea

The year 2017 continues to be an exciting year in Solar System exploration with eight missions currently active at Mars and more planned in the very near future. Please note that this article was written in spring 2017 and, as all the missions mentioned are active, some details may change for operational reasons. Venus and Jupiter both have an orbiting spacecraft whilst exploration at dwarf planet Ceres enters its final phase. The year sees three missions en route to their targets – Osiris Rex to asteroid Bennu, New Horizons to a January 2019 encounter with Kuiper Belt object 2014 MU69 and Hayabusa 2 to asteroid 162173 Ryugu. Sadly, 2017 saw the end of one of the most successful planetary missions of all time: Cassini at Saturn.

Hunt for Trojans

The OSIRIS-REx or Origins, Spectral Interpretation, Resource Identification, Security, Regolith Explorer (NASA loves acronyms, which is why NASA should stand for Never A Simple Acronym) is an asteroid study and sample return mission. It was launched on 8 September 2016 toward the asteroid 101955 Bennu. To enable the spacecraft to successfully match the orbit of Bennu the spacecraft will return to Earth on 22 September 2017 in order to use the Earth's gravity to adjust the orbit of OSIRIS-REx so that rendezvous with Bennu can occur in August 2018. During its cruise phase, OSIRIS-REx was used to search for near-Earth objects, known as Earth-Trojan asteroids. Trojan asteroids are trapped in stable gravity wells, called Lagrange points, which precede or follow a planet. OSIRIS-REx passed through Earth's L4 Lagrange point in February 2017. L4 is located 60 degrees ahead in Earth's orbit around the sun, about 150 million kilometres (90 million miles) from our planet. The search took place from 9 to 20 February 2017. Although no Earth-Trojans were discovered, the spacecraft's camera operated flawlessly and demonstrated that it could image objects two magnitudes dimmer than originally expected. During the search phase the camera onboard the spacecraft imaged a total of 17 main belt asteroids as well as Jupiter and some of its moons.

Juno at Jupiter

NASA's JUNO spacecraft arrived at Jupiter on 4 July 2016 after a 5-year cruise. The main propulsion system correctly placed JUNO into an elliptical polar orbit with a period of 53 days. After two of these orbits it was intended to ignite the propulsion system again to reduce the 53-day orbit into a 14-day science orbit. Difficulties with the propulsion system deferred this manoeuvre into 2017. In the middle of February 2017 engineers decided to leave the spacecraft in its current orbit. This will have no effect on the science return – it will just take a little longer. The current orbit brings JUNO to within 4,100 kilometres (2,600 miles) of the cloud tops and at its furthest will be 8.1 million kilometres (5 million miles) from Jupiter. Although science return will take longer, the current orbit does have certain advantages. Jupiter has intense radiation which can be very damaging. The 53-day orbit means JUNO will spend less time in this hazardous zone. JUNO will now be able to explore the far magnetotail, the southern magnetosphere, and the magnetospheric boundary region called the magnetopause. Results from these early orbits indicate that Jupiter's magnetic fields and aurora are bigger and more powerful than originally thought. The belts and zones in Jupiter's cloud tops so familiar to observers extend deep into

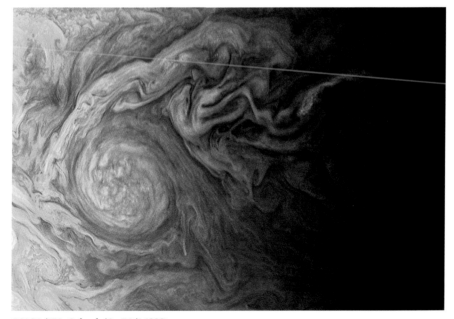

(NASA/JPL-Caltech/SwRI/MSSS)

the planet's interior. Using the JunoCam, the image on the previous page of Jupiter's atmosphere was taken on 2 February 2017 from an altitude of about 14,500 kilometres (9,000 miles) above the giant planet's swirling cloud tops.

Onward to the Kuiper Belt

After a highly successful flyby of the Pluto / Charon system by the New Horizons spacecraft on 14 July 2015, the mission team was granted a funding extension to enable New Horizons to explore potential targets within the Kuiper Belt. This funding extends the mission to 2021. Several observatories participated in a search for suitable candidates under the Ice Hunters project. On 26 June 2014, The Hubble Space Telescope observed a Kuiper Belt Object (KBO) in a part of the sky that New Horizons was travelling towards, the constellation Sagittarius. Located over 1.6 billion kilometres (1 billion miles) beyond Pluto, the KBO, now designated 2014 MU69 was officially chosen as the next objective for New Horizons in August 2015. During October and November 2015, a series of small thruster firings altered the trajectory of New Horizons toward this KBO. Flyby of 2014 MU69 (estimated diameter of between 20 and 40 km) will be on 1 January 2019. On 3 April 2017, New Horizons passed the halfway point to 2014 MU69. On that date, the spacecraft was 782.45 million kilometres (486,192,000 miles) from the KBO and the same distance from Pluto. The trajectory of New Horizons toward 2014 MU69 is shown here.

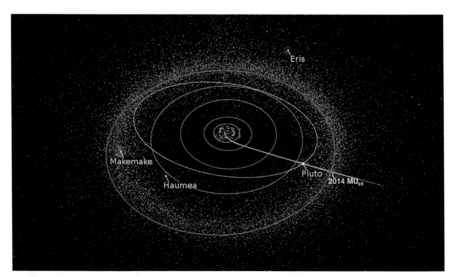

(NASA/Johns Hopkins University Applied Physics Laboratory/Southwest Research Institute/ Alex Parker)

Asteroid Explorer

Hayabusa 2 is an asteroid sample return mission launched by JAXA, the Japan Aerospace Exploration Agency on 3 December 2014 toward asteroid 162173 Ryugu. This mission continues the work of Hayabusa (English translation Peregrine Falcon) that returned a sample of the asteroid 25143 Itokawa on 13 June 2010. Hayabusa 2 returned to Earth in December 2015 for a gravity assist. The increase in velocity from this flyby shaped the orbit for an encounter with Ryugu in June 2018. It is scheduled to stay in orbit for one and half years before leaving the asteroid at the end of 2019 and returning to Earth around the end of 2020. An artist's impression of Hayabusa 2 collecting samples of Ryugu can be seen in this image.

(JAXA)

Aerobraking at Mars

The Trace Gas Orbiter (TGO) is a joint mission between the European Space Agency (ESA) and Russia's Roscosmos. The Orbiter was launched on a Russian Proton launcher from Baikonur Cosmodrome in Kazakhstan and carried with it the Entry, Descent and Landing Demonstrator Module (EDM) called Schiaparelli.

Although the lander did not survive its soft-landing attempt, it nevertheless provided valuable data for ESA's next lander. During 2017 ESA ground controllers need to re-shape the orbit of TGO from the highly elliptical orbit it is currently in to a more circular orbit so science operations can begin. In an ideal world, and with a bottomless budget, the spacecraft would have large fuel tanks so that the onboard propulsion system could be used to circularize the orbit in only a few weeks. However, to carry sufficient fuel for this manoeuvre, TGO would be too heavy for any current launcher to lift it. The cost would also blow ESA's science budget for years. Larger fuel tanks cost money, although Mars has something that engineers can use ... and it is free: an atmosphere.

On arrival at Mars, TGO was placed into a 250km x 98,000km orbit with a period of 4.2 days. It is the high point in the orbit, or apoareion, that needs to be reduced. By slowly lowering the lowest point in the orbit or periareion into the thin Martian atmosphere, atmospheric drag would rob TGO of energy so that the next apoareion would be lower. In early 2017 ground controllers conducted a series of manoeuvres using the onboard propulsion to reduce the apoareion to 34,000 km with a period of 24 hours and increase the orbital inclination to 74° to the equator. The atmospheric drag, or more correctly aerobraking phase, began on 15 March 2017 with a series of thruster firings to slowly lower the orbital low point from 250km to just over 100km. As atmospheric density increases and decreases over time it is not possible to state beforehand what is the optimum periareion distance. The ground controllers will 'feel' their way down into the upper atmosphere so as not to exert too large a force on the solar generating panels that will absorb most of the heat from atmospheric drag. A diagram showing the passes through the upper atmosphere resulting in a lowering of the apoareion is shown here. This aerobraking phase will last about a year and should be essentially complete by spring 2018 when the onboard propulsion system will complete the circularization process and leave TGO in a 400km circular science orbit, the image on the next page depicting TGO in orbit above Mars.

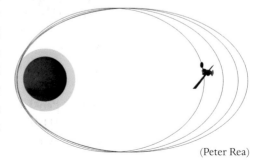

(Peter Rea)

(ESA)

Mars Summary

Mars continues to dominate Solar System exploration. Many are long lived missions and in the table below you can see how long the current Mars missions will have been at Mars as of 1 January 2018.

MISSION	ORIGIN	ARRIVAL DATE	TIME AT MARS ON 1 JAN 2018
Mars Odyssey	USA	24 October 2001	16 years 2 months 1 week
Mars Express	Europe	25 December 2003	14 years 1 week
Mars Exploration Rover-B	USA	25 January 2004	13 years 11 months 1 week
Mars Reconnaissance Orbiter	USA	10 March 2006	11 years 9 months 3 weeks 1 day
Mars Science Laboratory	USA	6 August 2012	5 years 4 months 3 weeks 4 days
Maven	USA	22 September 2014	3 years 3 months 1 week 2 days
Mangalyaan	India	24 September 2014	3 years 3 months 1 week
Trace Gas Orbiter	Europe	19 October 2016	1 year 2 months 1 week 5 days

Memories of a Saturn Orbiter

On 15 October 1997 Elton John was at number 1 in the charts with 'Candle in the Wind', a tribute to the recently deceased Diana Princess of Wales. Petrol prices in the UK were around 60 pence per litre and Britain's Andy Green set a jet-powered car record of 763.035 mph. The year 1997 also witnessed the launch of the Cassini / Huygens spacecraft to Saturn. I was at the southern end

(Peter Rea)

of Cape Canaveral Air Force Station to see this launch with my family. We had been there on the 13th for the first launch attempt but high winds forced a 48-hour postponement. On the early morning of the 15th the winds were much calmer. At precisely 04:43 am EDT the Titan 4 Centaur lifted off in complete silence. The launch pad was about 9 miles to the north. It therefore took about 45 seconds for the sound waves to reach me, by which time the launch vehicle was passing through some scattered cloud. My own photo of the launch vehicle passing through this cloud can be seen above. Little did I know then just how successful this mission would be or what discoveries were to come.

The origins of this Saturn mission go back to the early 1980s. In 1982, the American and European teams got together to discuss cooperation. A joint Saturn Orbiter and Titan Probe was suggested. This led to a proposal of NASA using a Mariner Mark 2 spacecraft as the orbiter with ESA building a probe to descend through the atmosphere of Titan. Originally called the Saturn Orbiter Titan Probe (SOTP) mission it was later called Cassini / Huygens after the Italian astronomer Giovanni Domenico Cassini and the Dutch mathematician and astronomer Christiaan Huygens. Cassini discovered four satellites of Saturn (Iapetus in 1671; Rhea in 1672; and Tethys and Dione in 1684) and in 1675 the

Cassini Division between the A and B rings of Saturn. As well as discovering Saturn's largest satellite Titan in 1655, Huygens also proposed that Saturn was surrounded by a solid ring.

Cassini was the 5 tonne Saturn orbiter built in the USA. Huygens was the European built atmospheric entry probe that would descend through the thick atmosphere of Saturn's moon Titan. Getting to Saturn was not easy. Even the powerful Titan 4 launch vehicle could not send the spacecraft directly to Saturn. There was a shortfall in energy that had to be gained from somewhere. That energy was gained by close flybys of Venus in 1998 and 1999 followed by a return to Earth in 1999. The spacecraft now had enough energy to fly past Jupiter on 30 December 2000. The additional increase in velocity from this flyby sent Cassini / Huygens on a course that would intercept Saturn on 1 July 2004. The Huygens probe released from the orbiter on 25 December 2004 and entered the atmosphere of Titan on 14 January 2005. Over the next 12 years the Cassini orbiter would make repeated flybys of Titan and a host of other moons. It would explore all regions of the Saturn system, this diagram showing all the orbits of Cassini around the planet. Note that some of these orbits took Cassini

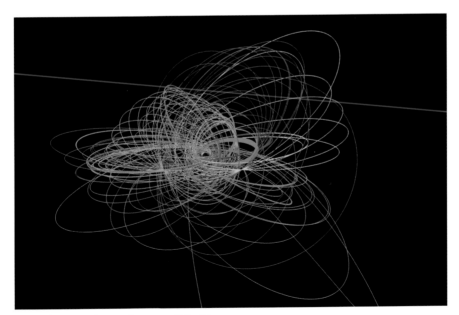

(NASA/Jet Propulsion Laboratory-Caltech)

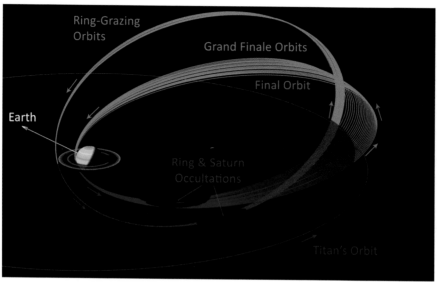

(NASA/Jet Propulsion Laboratory-Caltech)

over the polar regions for dramatic views. However, all good things must come to an end, and 2017 should have seen the final days of Cassini, as it will have been intentionally crashed into Saturn. This disposal of Cassini is necessary as onboard fuel for the propulsion system is running low, and in accordance with planetary quarantine procedures Cassini must not be allowed to crash into any of the moons because of possible contamination. In the final months of the mission, Cassini will make some daring passes between the innermost ring and the planet's atmosphere. These final passes are shown above. The series of passes through the gap began in April 2017 and was scheduled to end with the deliberate crashing of Cassini into the atmosphere of Saturn on 15 September 2017. The artist's impression on the next page depicts Cassini passing between the rings and the atmosphere of Saturn.

If I had to pick one discovery that stands out above all others it would be the possibility of life on one of Saturn's moons. Enceladus (discovered by William Herschel in 1789) is an active moon that hides a global ocean of liquid salty water beneath its crust. Scientists discovered evidence of Enceladus' internal ocean from gravity measurements based on the Doppler effect and the magnitude of the moon's very slight wobble as it orbits Saturn. The data

(NASA/Jet Propulsion Laboratory-Caltech)

was consistent with the existence of a large global ocean inside the moon. The measurements suggested a large sea about 10 kilometres (6 miles) deep beneath an ice shell about 30 and 40 kilometres (19 and 25 miles) thick in the southern polar region. On many close flybys of Enceladus, Cassini has flown through and sampled the waters from this subsurface ocean that is being jettisoned into space, this illustration depicting Cassini diving through the Enceladus plume in 2015. The material shoots out at around 1,200 kilometres per hour and forms a plume that extends hundreds of kilometres into space. Some of the material falls back onto Enceladus and some escapes to form Saturn's vast E ring.

(NASA / JPL-Caltech)

Cassini's chemistry analysis strongly suggests the Enceladean seafloor has hot fluid vents – places that on Earth are known to teem with life, though no guarantee that is happening on this moon. This possibility of life on another body other than Earth may well lead to a follow up mission to Saturn and Enceladus at some point in the future. Europa, one of the Galilean moons of Jupiter also shows evidence of sub surface water and is the target of future missions from NASA and the European Space Agency.

The Cassini / Huygens mission was first mooted as a possible joint NASA / ESA project back in 1982 and 35 years have elapsed from those first exploratory meetings to the end of the mission in 2017. I was there at the start, and I shall have watched with much admiration as the mission came to an end just short of 20 years after launch. Missions like Cassini / Huygens do not come around very often. It was a privilege to be there at the beginning and the end.

Solar System exploration continues to excite and inspire, and next year promises to be no different.

Anniversaries in 2018

Neil Haggath

Kepler's Laws

This year sees the 400th anniversary of the publication of Kepler's Laws of Planetary Motion, which were a forerunner to Newton's Law of Universal Gravitation, and the key to establishing the scale of the Solar System.

Johannes Kepler (1571–1630) is often said to have been the world's first astrophysicist. While his contemporary Galileo established the methods of what we now call experimental physics, Kepler was the first to realise that the motions of astronomical bodies could be described by the laws of physics and mathematics. He was an early supporter of Copernicanism – but as Copernicus himself admitted, there were anomalies in the observed motions of the planets which his

Portrait of Johannes Kepler in around 1610 by an unknown artist. (*Wikipedia*)

heliocentric theory could not explain – that is, discrepancies between their observed positions and those predicted by his model.

Kepler first proposed a model of the structure of the Solar System, involving the regular solids of classical Greek geometry, which was frankly silly, and had more to do with ancient Greek mysticism than with science. But he soon realised that it did not work and, in the manner of a true scientist, abandoned it and started again.

In 1600, Kepler went to Prague to work as assistant to Tycho Brahe (1546–1601). Tycho was perhaps the most meticulous observer of the pre-telescopic era (what a tragic irony that he died just seven years before the invention of the telescope) but was no theorist, and refused to accept Copernicanism. Kepler, in contrast, was not much of an observer, but a brilliant mathematician and theorist. After Tycho's death in 1601, Kepler inherited all his observations of the planets – the most accurate to date – and set about trying to reconcile them with Copernican theory.

He eventually realised that the discrepancies could be explained by assuming that the planets' orbits were not circular, as Copernicus had assumed, but elliptical; by a painstaking process of trial and error, trying ellipses of different eccentricities, he found the best fit for each planet.

In 1609, he published the first two of his Laws of Planetary Motion. Then in 1618, he hit upon his third and most important law, which states that the square of a planet's orbital period is proportional to the cube of its mean orbital radius. As well as being a forerunner of the inverse square law of gravity, this relationship proved vital in determining the distances of the planets; as their orbital periods could easily be measured, it was only necessary to measure the distance of the Earth from the Sun, and those of all the other planets could be calculated.

Newton's Telescope

350 years ago, Sir Isaac Newton (1642–1726) made the first reflecting telescope. He was not the first person to realise that a telescope could be made using a mirror to focus light; the Scottish mathematician James Gregory described the principle five years earlier, but by his own admission, lacked the practical skills to make such an instrument. Newton was the first to actually make one, in 1668; his first reflector was a tiny instrument of just 1.5 inches aperture. He presented it to the Royal Society in 1672; the Society's interest later led him to write *Opticks*, his next most famous work after the *Principia*.

Gregory's design was similar to what we now call a Cassegrain, with a convex secondary mirror directing light through a hole in the primary. Newton, however, pioneered a simpler design, using a flat diagonal secondary mirror and the eyepiece at the side of the main tube. This design is still known as a Newtonian reflector.

Rev William Rutter Dawes (1799–1868)

15 February is the 150th anniversary of the death of Rev. William Rutter Dawes, best known to amateur astronomers as the man who gave his name to the 'Dawes Limit', a measure of the resolving power of telescopes.

Dawes was a renowned observer, who became known as 'Eagle Eye', due to his keen eyesight. He became a friend of such luminaries as Sir John Herschel and William Lassell. He owned a 3.8-inch Dollond refractor, and was later employed to run the private observatory of the wealthy George Bishop, where he had the use of a 7-inch instrument.

Dawes had a particular talent for observing double stars. In 1867, he published a paper in the *Memoirs of the Royal Astronomical Society*, in which he presented a formula for calculating the closest separation at which a double star could be resolved by a telescope of a given aperture. His criterion for 'resolving' a double star was that the Airy disks of the two components could be distinguished, though they might still overlap.

William Rutter Dawes. (*Wikipedia*)

His formula states that for two stars of magnitude 6, using a refracting telescope of aperture 'a' in inches, the closest separation at which they can be resolved is (4.56 / a) arc seconds. This is known as the Dawes Limit.

Today, the Dawes Limit is often quoted as a general purpose formula for the resolving power of any telescope. However, it should be remembered that it was intended for the very specific case of splitting double stars with a refractor.

George Ellery Hale (1868–1938)

29 June is the 150th anniversary of the birth of George Ellery Hale, the man best known for building the world's biggest telescope, not just once, but twice.

Hale excelled at almost every aspect of professional astronomy – as an observer, a theorist and a builder of instruments. While an undergraduate, he invented the spectroheliograph, an instrument for observing the Sun in light of a single wavelength. In 1908, he used the Zeeman Effect – the splitting of spectral lines by a magnetic field – to establish that sunspots are magnetic features; he later discovered the symmetry of their polarity in opposite hemispheres, and that their polarity is reversed in each 11-year solar cycle.

George Ellery Hale. (*Wikipedia*)

However, Hale is best remembered as a founder of observatories and a builder of great telescopes. In 1897, he founded Yerkes Observatory in Wisconsin, and became its first Director. Then, as now, many observatories were funded by donations from wealthy individuals; Yerkes was named in honour of its benefactor, Charles T. Yerkes. It was equipped with a 40-inch refractor, which to this day remains the biggest refractor ever built – a record which will never be beaten, as this is just about the physical limit to how large a lens can be made, without deforming under its own weight.

Hale soon realised that the future lay with reflecting telescopes, and also recognised the advantages of mountaintop sites. In 1904, he founded Mt. Wilson Observatory in California. Its first large telescope was a 60-inch reflector, which was at the time the world's biggest operational telescope; the 72-inch 'Leviathan of Parsonstown', built by the Third Earl of Rosse in 1845, had long since fallen into disuse and ruin. In 1917, the 100-inch Hooker Reflector was completed – named for its benefactor, John D. Hooker – which was the first to surpass the Leviathan in size. Edwin Hubble would later use the telescope to measure the redshifts of galaxies, and discover the expansion of the Universe. It remained the world's biggest for the next three decades.

Yet Hale was still not satisfied, and was determined to build a telescope twice the size of the Hooker! Work on Palomar Mountain Observatory, also in California, was begun in 1928 – but there were long delays, and its 200-inch reflector was not completed until twenty years later. Sadly, Hale did not live to see the completion of his greatest project, as he died in 1938. The 200-inch was rightly named in his honour.

Among his many skills, Hale possessed a remarkable talent for persuading millionaires to bankroll his projects. Legend has it that he once switched the place cards at a dinner party, in order to seat himself alongside his next intended 'target' – and before the evening was over, the deal was agreed!

Henrietta Swan Leavitt (1868–1921)

This year also sees, on 4 July, the 150th anniversary of the birth of another pioneering astronomer – Henrietta Swan Leavitt, who played a vital part in establishing the cosmic distance scale.

From 1893, Leavitt worked at Harvard College Observatory as a 'computer' – which in those days meant a person who performed laborious calculations by hand, to process the data collected by the observers. The Observatory's Director, Edward Pickering, employed exclusively women for this purpose, and they became known as 'Pickering's Harem'. Part of Leavitt's task was to measure and catalogue star brightnesses on photographic plates, and she was assigned to study variable stars.

Henrietta Swan Leavitt. (*Wikipedia*)

Between 1908 and 1912, she made the discovery which assured her of a place in astronomical history. She studied a class of variable stars known as Cepheids; these all vary with very regular periods, but the length of their periods varies greatly between different stars. Leavitt studied Cepheids in the Magellanic Clouds, the two small satellite galaxies of the Milky Way. At the time, the distances of the Magellanic Clouds were not known with any certainty, but they were known to be small in size in comparison to their distances; therefore, all the stars within each one could be regarded as being at roughly the same distance from us. This in turn meant that their apparent brightnesses directly corresponded to their intrinsic luminosities.

Due to this correspondence, Leavitt discovered that the periods of Cepheids are directly related to their luminosities – the more luminous the star, the longer its period. She published what became known as the Period-Luminosity Relationship in 1912. The relationship had far-reaching consequences; if the distances of a few Cepheids could be established, then the relationship could be quantified. Just two years later, Henry Norris Russell published what became known as the Hertzsprung-Russell Diagram, which provided a method of determining the distances of stars which were too far away to be directly measured by parallax. This enabled the distances of the closest Cepheids to be measured. Thereafter, the distance of any Cepheid could be determined, simply by measuring its period of pulsation; the period gave its intrinsic luminosity, and comparing that with its apparent brightness gave its distance. Cepheids therefore became the first known 'standard candle'.

Cepheids are so extremely luminous that individual ones can be identified in other galaxies, at distances of many millions of light years. Edwin Hubble later used them to determine the distance of the Andromeda Galaxy, thereby proving that the 'spiral nebulae' were other galaxies, and not smaller structures within our own, as some believed. Leavitt had therefore established the first method of measuring distances beyond our Galaxy.

Hubble himself said that Leavitt deserved a Nobel Prize for her discovery. She was in fact nominated for that ultimate honour – but sadly, not until three years after her premature death, by someone who was not aware that she had died. The Prize is not awarded posthumously.

Apollo 8

December sees the 50th anniversary of Apollo 8, the first manned space mission to leave Earth orbit and fly to the Moon.

Apollo 8 was the second manned Apollo mission, and the first to be launched by the huge Saturn V rocket. It was part of NASA's step-by-step build-up towards

Taken in December 1968 by astronaut Bill Anders during the Apollo 8 mission, 'Earthrise' is a photograph of the Earth rising above the desolate lunar surface. The image has been described as being the most influential environmental photograph ever taken. (*NASA / Bill Anders*)

the first Moon landing the following year. The mission was originally intended to be a test of the complete Apollo spacecraft, including the Lunar Module, in Earth orbit, with Apollo 9 the first to go to the Moon. However, delays in the development of the Lunar Module, together with reports that the Soviet Union was planning a manned flight around the Moon, led to the two missions being swapped. Apollo 8 did not carry a Lunar Module, but the Command and Service Module alone went to and orbited the Moon.

The crew were Frank Borman (Commander), Jim Lovell and Bill Anders. Borman and Lovell had previously flown together on Gemini 7, and Lovell had also commanded Gemini 12; Anders was making his first and only spaceflight. Lovell, of course, would later command the ill-fated Apollo 13. They became the first humans ever to see the Moon's far side with their own eyes.

Apollo 8 was launched on 21 December 1968, and returned to Earth on 27 December, after a flight time of six days and three hours. After a three-day outward flight, the spacecraft entered lunar orbit on Christmas Eve, and orbited the Moon ten times, at a height of just over 100 km, in the next 21 hours.

During the fourth orbit, Anders took the famous 'Earthrise' photograph, showing the Earth rising above the lunar landscape, which became one of the most iconic images of the Space Age. It has been said that the publication of the image, more than any other single event, raised public awareness of ecological and environmental concerns, as it suggested the isolation and fragility of our world. In this respect, the image has been described as the most important legacy of the space programme.

Comets and How to Photograph Them

Damian Peach

Introduction

Over the last several years we have been treated to some dramatic naked eye comets gracing our skies. Comets C/2006 P1McNaught and C/2011 W3 Lovejoy were wonderful sights to those located in the southern hemisphere, though sadly neither were favourably placed for northern hemisphere observers. Comet ISON in 2013 promised a reprieve from this northern drought but sadly it disintegrated near perihelion. C/2014 Q2 briefly reached 3rd magnitude in 2015 but the wait continues for a really bright comet to grace northern skies (such as Hale Bopp or Hyakutake did back in 1996/97).

Comet C/2014 Q2 (Lovejoy) on 20 January 2015. This comet has been one of the finest in recent years, briefly reaching 3rd magnitude and sporting a spectacular ion tail. This image shows it near its peak, with a multitude of streamers within its tail. (Image D. Peach)

However, when such a comet does again return to northern skies we have never been more ready. With digital camera technology now having replaced film photography we have never been better equipped to capture dramatic shots of these brief visitors to our Solar System when they appear. The article describes some of the techniques to employ in how to capture good images of these fleeting visitors.

Dirty Snowballs

So what exactly are comets? Where do they come from? Why do they often have that characteristic tail? Comets are small bodies of ice, rock, dust and frozen gases ranging in size from just a few hundred meters to as large as over 40 kilometres. They are thought to originate from the distant reaches of our solar system where the Scattered disc and Oort cloud are located – regions of icy material left over from the formation of the solar system. Current thinking suggests that shorter period comets originate from the scattered disc while long period comets originate from the Oort cloud.

Gravitational perturbations from the giant outer planets are thought to cause these distant bodies to be sent onto a course toward the inner solar system. As they slowly draw closer to the Sun, solar heating causes volatile materials within the comet to vaporize carrying dust away with them, forming a large tenuous atmosphere around the comet called the coma. Pressure from solar radiation and the solar wind shape this into a tail of dusty material pointing away from the Sun. This dust tail is illuminated by the Sun, but a second tail also forms from the gases released from the comet. This is often called the gas or ion tail which glows due to ionisation. These tails often point in slightly different directions from each other.

The dust tail is left behind along the comets orbital path while the ion tail, being more strongly affected by the solar wind, always points directly away from the Sun as well as often presenting a notable bluish colour – in stark contrast to the hazy white dust tail. Another type of tail sometimes seen with bright comets is the anti-tail. This tail can be seen emanating from the nucleus and seemingly pointing toward the direction of the Sun. This appears when the Earth passes through the plane of the comet's orbit, so you see lots of edge-on dust as a Sunward pointing spike. In recent times Comet Lulin had a near-permanent anti-tail as its orbital inclination was within 2 degrees of Earth's orbit (the ecliptic).

Comets typically come in two general types known as short and long period. Short period comets have an orbital period around the Sun of less than approximately 200 years while long period ones can have orbital periods of thousands or millions of years. For example Halley's Comet orbits the Sun every 75 or 76 years and when at its furthest from the Sun is located just beyond the orbit of Neptune, making this a short period comet. However, long period comets can have extremely long orbital periods. Comet Hale-Bopp for example, which I am sure many readers will remember passed through the inner solar system in 1997, will not return again for well over 2000 years, its orbit taking it into the outer reaches of the solar system (so I am afraid if you missed it back in 1997 it is going to a very long wait until it comes back again)!

A third type of comet, known as the Sungrazer, is responsible for many of the most spectacular comets throughout history. These comets are characterised by their very close perihelion distance (meaning they pass very close to the

Comet Hale-Bopp was one of the most widely observed comets in living memory. In this photograph, taken during its close pass in 1997, its white dust tail and blue ion tail are clearly defined. (Image M. Mobberley)

Sun) which can result in them appearing extremely bright and spectacular for a short time. There are a few groups that comprise this type of comet, the most well-known of which are the Kreutz group of sun grazing comets. These are all thought to come from the same parent body which fragmented many centuries ago. Many great comets throughout history were sungrazing comets (Ikeya-Seki, West, McNaught and Lovejoy being good examples). Comet ISON was also a sungrazer and passed very close to the Sun in late 2013, briefly appearing as an impressive object before disintegrating.

Photographing Brighter Comets

Unlike many forms of astrophotography, which can be quite costly and complex, shooting bright comets can be a surprisingly straightforward yet greatly rewarding pursuit. With a few points kept in mind you can start capturing spectacular images quite easily.

For bright comets (which we can define as visible to the naked eye) a digital SLR equipped with a standard 18–55mm lens and mounted on a sturdy tripod

Comet 17P Holmes, captured by the author back on 12 December 2007 using a digital SLR attached to a small 80mm telescope. 20 x 60-second exposures were summed together to make the final image. Even modest equipment can work very nicely with bright comets. (Image D. Peach)

will suffice. Firstly you need to operate the camera in 'M' mode giving you full control over the camera settings. Carefully focus the camera on a bright star or the moon if available (or even the comet itself if it is sufficiently bright.) Many DSLRs today are equipped with a 'live view' mode which allows you to focus a magnified live image on the cameras LCD display which is extremely useful for accurate focusing.

You will want to set the lens aperture to its widest value (lowest F-number) and also to set the ISO sensitivity to around 400 to 800, though for fainter comets higher ISOs in the 1600 to 3200 range may yield better results (although more noisy).

A sturdy tripod is, of course, a must for any form of astrophotography. You will want a remote release to trigger the camera shutter without having to actually touch the camera. However, many DSLRs these days have various modes that allow a time delay series to be taken, so you can trigger the exposure by pressing the shutter release button and the camera will wait a few seconds before actually triggering the shutter. I have used this mode myself on many occasions with great success.

Experiment with exposure times. There is no firm guide and it will depend on how bright the comet is and also how bright the sky is (brighter comets are often to be found in twilight skies). In general however you will always want to keep single exposures below around 15 seconds to avoid trailing, and the higher the focal length of lens you use the shorter amount of time you have to avoid trailing due to Earth's rotation. Of course to avoid this issue you can mount a camera on a small driven mount, although many may not have anything suitable.

Do not just shoot a few single shots but take a series of images that you can stack together to smooth out noise (especially important if using higher ISO settings). There are many programs available that can easily stack together images. The planetary image processing program Registax will do a good job in stacking images from SLR cameras and is very easy to use. Deep Sky Stacker also works well and both programs are freely available for download.

CCD Imaging through the Telescope
Of course with a telescope equipped with a CCD camera a whole range of comets become available, not just those rare spectacular naked eye targets.

At any one time there can be hundreds of comets within reach of amateur telescopes equipped with CCD cameras. Though most are of course very faint, there can often be a few around at any one time that make excellent targets for small telescopes.

Pretty much any telescope can be used to shoot comets though fast refractors, Newtonians and Ritchey–Chrétiens are often the favoured choice of observers for their wide flat fields of view and fast f-ratios. Digital SLRs can again be employed, mounted to the telescope, and will certainly produce good results, though many observers also use dedicated astronomical CCD cameras such as those manufactured by companies like SBIG and FLI. These types of cameras are certainly not cheap but offer the highest image quality. Using dedicated CCD cameras will require the usual processes employed with deep sky work, such as applying dark frames and flat fields to fully calibrate the images before further processing is applied.

In general when imaging comets through a telescope you will immediately notice the comet is moving with regard to the background star field. Some comets move much faster than others but typically any single exposure longer than 4 to 5 minutes will result in the comet itself trailing in the image. The easy solution is to take several short exposures, such as 5 x 60-second exposures and then stack the resulting images on the comet itself. This of course will result in a nice sharp comet, albeit with a trailed star field, although this method is the one employed by most comet observers. More complex image processing routines allow combining two images – one stacked on the star field and the other stacked on the comet, but the scope of describing this process is beyond an article such as this.

Another great way to photograph comets is by using remote telescopes. These telescopes have the benefit of being located under extremely dark skies and represent a great solution for those stuck under cloudy or light polluted skies. For a modest fee you can log onto a telescope online and control it in real time to target the object of your choice. Real time views of the sky and weather are provided and you can view images from the telescope as they are taken and download your data in the usual FITS format for processing. The real benefit here of course is that this enables you to use telescopes located at top quality dark sky locations and also make useful observations when nothing may be possible from home. Today many telescopes are operated in this manner (both amateur and professional) with great success.

Two rather less bright but lovely comets that have graced our skies recently have been C/2015 V2 (Johnson) and C/2015 ER61 (PanSTARRS.) Though neither comet was especially bright, both were fine objects for small telescopes. (Images D. Peach)

Comets in 2018

Forecasting comets can be a lot like forecasting the weather – rife with uncertainty! At the time of writing the following objects should hopefully become worthwhile targets for small telescopes.

COMET	PERIHELION DATE	PEAK MAGNITUDE
21P/Giacobini-Zinner	10 September 2018	5
46P/Wirtanen	12 December 2018	3

Of course it is quite possible other comets will be discovered, and which will brighten quickly, or an existing object will undergo an outburst. Just recently, another Lovejoy comet discovery brightened rapidly to become a nice object for a few weeks before disintegrating near perihelion. This is one of the great

C/2017 E4 (Lovejoy) seen here near its best on 5 April 2017. Discovered less a month earlier, it briefly produced some nice activity, but sadly disintegrated close to perihelion. A prime example of the unpredictable nature of comets! (Image D. Peach)

(and frustrating) things about comet observing – you never know what is going happen, and it is perhaps for that reason comets remain such captivating objects to observe and photograph.

Further Reading and Information

There are many good sources going into much greater detail than has been possible here, and here you will find some excellent sources of information that I can personally recommend, including the book *Hunting and Imaging Comets* by Martin Mobberley.

Websites

For all the very latest information on current comets by Seiichi Yoshida visit **www.aerith.net**

Minor Planet Centre Observable Comets: **www.minorplanetcenter.net/iau/ Ephemerides/Comets**

For remote observing visit the world's foremost remote telescope facility: **www.itelescope.net**

Double and Multiple Stars

John McCue

Our sun is an only child, but most of the two hundred billion stars in our Milky Way galaxy are in families of two or more. But beware: if you think you have spotted one of these groupings through a telescope, you could be mistaken. Seeing two stars close together could be nothing more than an accidental line-up and what you may be looking at is an optical double.

Unlike optical doubles, binary systems are made up of two stars which are gravitationally associated and orbit their common centre of mass, or barycentre, their very nature meaning that their separation and position angle are subject to change over a period of time. Perceived changes in separation arise through a combination of their actual distance apart and the angle from which we view the binary. Binary systems are much more useful than optical doubles, presenting the only direct way of measuring the mass of a star.

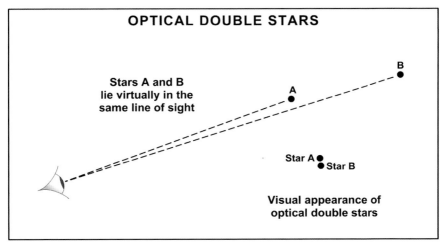

OPTICAL DOUBLE STARS

Stars A and B lie virtually in the same line of sight

A
B

Star A ●
● Star B

Visual appearance of optical double stars

Optical doubles consist of two stars which happen to lie more or less in the same line of sight as seen from Earth and which therefore only appear to lie close to each other when we see them in the sky. In reality, one star may lie many times further away than its 'companion', as can be seen in this diagram. (*Brian Jones / Garfield Blackmore*)

The amateur observer can play a big part in this mass determination by measuring the relative position of the two stars, the two parameters that quantify this being the position angle and separation. Watching binaries perform their celestial waltzes over time (sometimes very many years) enables astronomers to measure the orbital period. Stars hurtling round each other in less than a year have been seen, while some take as long as thousands of years.

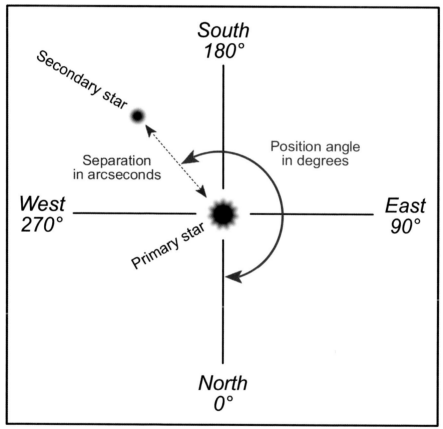

The position angle is the angle measured through east from an imaginary line extending north from the brighter component. Knowing the position angle of the binary, together with the angular separation of the individual stars, will help to determine in what direction from the brighter component to look for the fainter companion as well as giving an idea of the visual appearance of the star prior to observation. It must be borne in mind that the directions shown here relate to the image that would be seen through an inverting telescope, such as a Newtonian reflector. (*Brian Jones / Garfield Blackmore*)

By knowing the distance of the binary and how far apart they are as seen through a telescope, the actual size of the binary star's orbit can be worked out. Together with its orbital period, Kepler's Third Law can then be used to find the sum of the stars' masses.

Next, careful observations of the two stars against the matrix of background stars will show the sizes of the two individual orbits of the stars around their barycentre and so, like the balance of a big child and a small child on a see-saw, the individual masses of the stars can be found.

Now we know the masses of the stars, everything else about the stars makes sense because a star's mass determines its luminosity, its temperature profile, how long it will live, and which elements it can forge in its fiery core. Without stars to synthesise the elements of which we and our Earth are made, we would not be here!

Like odd couples, binary systems come in many varieties – little and large, bright and faint, and rainbow colours. The brightest star in the entire sky, Sirius (α Canis Majoris), well known in the northern winter / southern summer night skies, is a prime example. Sirius is around twice the size of our own Sun, but decidedly hotter, with a surface temperature of 10,000 K, as compared to the Sun's 6,000 K. Hence, Sirius is about 25 times as luminous as the Sun. This is bright, but not excessively so, and Sirius only appears so prominent because of its relative proximity, shining from a distance of just 8.6 light years. But Sirius has an odd companion, a white dwarf star with about the same mass as the sun. This object is the dying collapsed remnant of a once active star, the matter forming it being crammed into a sphere a little smaller than the Earth, making it so dense that a teaspoon of its material would weigh as much as two fully-laden transit vans.

This white dwarf companion is known as The Pup, in keeping with the fact that Sirius is often referred to as the Dog Star and, because of the glare of the brilliant primary star, splitting these two stars in a telescope is one of the most challenging tasks for a double star observer.

If you had applied to join the Roman army a couple of millennia ago, legend has it that you would have been asked to look at Mizar (ζ Ursae Majoris), the middle of the three stars in the 'handle' of the Plough, the well-known asterism which denotes the tail and hind quarters of Ursa Major (the Great Bear). If you could see a fainter star nearby, you were in! That other star is the 4th magnitude Alcor, located 11.8 minutes of arc from Mizar.

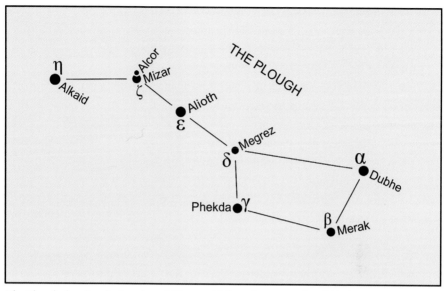

The Plough showing the location of Alcor and Mizar. (*Brian Jones/Garfield Blackmore*)

In 1617, the Italian astronomer Benedetto Castelli, a friend and student of Galileo, looked at Mizar and saw that it was really two stars, thus becoming the first astronomer to discover a double star through a telescope. More surprises were to come in the modern era. In 1889, Mizar became the first binary star to be discovered by spectroscopic means, following which Mizar's fainter component was also found to be a spectroscopic binary, meaning that Mizar consists of two sets of binaries – making it a quadruple star. Not only were Mizar A and Mizar B found to be spectroscopic binaries, but Alcor was too, making a total of six stars all together. But are they all a family? Measurements indicate that Alcor and Mizar lie at distances of around 82 light years and suggest that they are not too far apart, with a distance between them of perhaps as little as a quarter of a light year, making them part of a gravitational family. Mizar and Alcor could well indeed be six stars together in a merry group!

In your telescope, you will also see a fainter star midway, but off-line, between Mizar and Alcor. This star was first seen by the aforementioned Castelli, although this observation was seemingly overlooked by the German astronomer Johann Georg Liebknecht when he observed Mizar and Alcor telescopically in 1722. He believed this midway star to be a planet and named it after his patron Landgrave Ludwig, and even to this day it is often called Sidus Ludoviciana (Ludwig's Star).

Sometime later in the same century, the great William Herschel spent 25 years measuring the separation and position angles of Castor (α Geminorum), Algieba (γ Leonis), Izar (ε Boötis), Zeta (ζ) Herculis, Delta (δ) Serpentis and Porrima (γ Virginis) and became convinced that they were binaries. His son John Herschel carried on his father's work on double stars.

The first catalogue of double stars (which included binaries) was compiled in 1777/1778 by the Czech astronomer and Jesuit priest Christian Mayer, noted for his study of double stars. He published the catalogue in 1781, his astronomical enthusiasm having not been dented by the Pope's dissolution of the Jesuit order in 1773.

In 1827 the amazingly versatile German-born astronomer Friedrich Georg Wilhelm von Struve published his *Catalogus Novus Stellarum Duplicium et Multiplicium*. This was a tome of 3,112 double stars observed from the Dorpat Observatory, most of which he had discovered himself, this being followed by additions to the catalogue in 1837 and 1852, following which his son Otto Wilhelm Struve added to the total. In 1906, when the American astronomer Sherburne Wesley Burnham – a prominent discoverer of double

Johann Georg Liebknecht. (*Wikimedia Commons*)

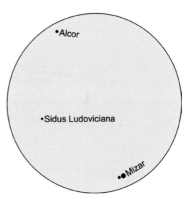

A detailed view of the Alcor and Mizar system showing the location of Sidus Ludoviciana. (*Brian Jones / Garfield Blackmore*)

stars – compiled and published *A General Catalogue of Double Stars within 121°* *of the North Pole*. In this catalogue, Burnham enumerated 13,665 double and multiple stars, and set the standard for a long time.

Burnham started a card catalogue of observations to keep the ball rolling, and this was maintained by the American binary star astronomer Robert Grant Aitken who found another 3,000 doubles from the Lick Observatory. Aitken employed Hamilton Moore Jeffers, who transferred all the handwritten cards to punched cards readable by computer. He then contacted Willem Hendrik van den Bos, and together they created a catalogue of nearly 65,000 doubles covering the whole sky.

Further computer-related advances were made to the data, and eventually it came under the maintenance and protection of the US Naval (National) Observatory, Washington DC, where observations from around the world are added nightly, and there are now details of 115,000 double star systems in the Washington Double Star Catalogue (WDS).

I was in New Zealand for the first time during their midwinter of 2016 and saw Alpha (α) Centauri for the first time. This star is only visible from latitudes south of around 30°N and its main claim to fame is being the closest naked eye star to the Sun, shining from a distance of 4.3 light years. Physically, it is very much like our Sun, though just a little larger and more luminous. It is a

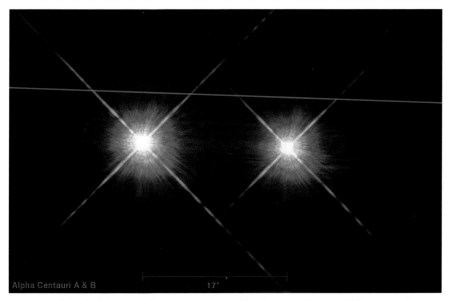

An image of the double star Alpha Centauri obtained by the NASA/ESA Hubble Space Telescope. (*ESA/Hubble & NASA*)

binary with yellow and orange magnitude 0 and 1.2 components which orbit each other over a period of 80 years. The orbit is quite elliptical, the distance between the two components varying between 11 AU and 35 AU. When at their farthest they are separated by 22 arc seconds and so the pair is easily visible in moderate telescopes. The yellow and orange hues of this pair make a marvellous contrasting sight in the telescope.

There are, of course, many such double star joys of colour contrasts and in the northern hemisphere, probably the finest being the blue and yellow Albireo (β Cygni).

The well known star Proxima Centauri is a faint 11th magnitude red dwarf located a little over 2° to the south west of the α Centauri pair. Why is this faint star so well known? As its name suggests, it is slightly nearer to us than α Centauri – by about 0.1 light years. No-one knows for certain whether these

This image of the optical double star 56 Andromedae, seen here at lower-left of picture, also contains the open galactic cluster NGC 752, which is situated slightly to the north west (upper right) of the double. The lining up of these objects is purely fortuitous, the cluster lying at a distance of around 1,300 light years as compared to the magnitude 5.8 and 6.1 stars forming the double which shine from distances of around 320 and 920 light years. The image was taken with a Canon 600D SLR camera on an 80mm diameter apochromatic refractor. (*Neil Webster*)

three stars are a family, but what was memorable for me was the news on my return to the UK of the discovery of an earth-sized planet going around Proxima, and that conditions on this planet may well be suitable for life.

As well as allowing us to directly measure stellar masses, binaries give us an accurate means to scale the universe out to the farthest galaxies. Supernovae of Type 1a are the outcome of a remarkable kind of binary system – one that contains a carbon white dwarf and another, younger, star in orbit around it. The strong gravity of the dense white dwarf (a result of the matter forming the star being squeezed into a very small volume) pulls material from the atmosphere of its partner. An accretion disc of swirling material forms around the white dwarf and gradually accumulates onto the surface of the dwarf, thereby increasing its mass. Gaining this extra mass is a serious problem for the white dwarf because its internal consistency is a strictly defined physical state known as electron degeneracy. This state of affairs comes about because its end-of-life collapse increases the internal pressure so much that the electrons surrounding the nuclei of the star's atoms become squashed into the lowest possible energy states, and this electron degeneracy pressure supports the star. The Indian astrophysicist Subrahmanyan Chandrasekhar then worked out that the optimum mass of a white dwarf is 1.4 times the mass of the Sun. Any more than this and the gravity of the extra mass would cause more crushing and the star would cease to be a white dwarf and go on to be a neutron star or even a black hole, depending on how much extra mass it has.

The white dwarf in the binary thus has a dilemma. It now has too much mass and begins to collapse and heat up, and its carbon begins to fuse to heavier elements. Being under so much pressure, the degenerate white dwarf cannot expand and cool down like a normal gas, so the rise in temperature runs away, until a point is reached when something has got to give. The temperature becomes so high that the material starts behaving like a normal gas again, resulting in a sudden explosive expansion – a supernova cataclysm – which destroys the white dwarf completely, scattering its elements into space. Here is the amazing surprise though – the strictly mathematical nature of this process means that the supernova explosion is always the same brightness, about ten billion suns. Not only that, but it has a recognisable spectrum and light curve, which means we can play a neat trick. Seeing one of these supernovae means we know how bright it really is, and seeing how bright it appears to be, we can work out its distance. This standard candle means we can work out the distance to the host galaxy and then, by measuring the redshift of the galaxy, we can study the expansion of the universe. It was precisely this technique that resulted in the astonishing discovery of the accelerating expansion of the universe, implying an unknown agent which has been termed dark energy.

Finally, mention must be made of an even more recent discovery – the first detection of Einstein's gravity waves. In 2016, a momentous announcement was made by astronomers at gravity wave detectors in Louisiana and Washington. A binary system, comprised of two black holes spiralling together at nearly the speed of light, collided and destroyed each other in a gigantic cosmic death wish. The two black holes swirled through the surrounding matrix of space-time causing gravitational waves, like two paddles surging through water and making waves.

Gravity waves are all around us, wherever a mass accelerates through space-time, but they are incredibly weak. It took the cataclysm of two black holes, maybe up to two billion light years away, to generate waves big enough to detect here on Earth.

Without binary systems, we would have struggled to make these ground-breaking discoveries.

Some Pioneering Lady Astronomers

Mike Frost

One of the inspirations of being director of the British Astronomical Association's Historical Section is that I follow in the footsteps of illustrious predecessors. Former directors include Colin Ronan, author of many books on astronomical subjects; Commander Derek Howse of the Royal Observatory Greenwich; and Ernest Beet, a past-president of the Association.

However, one of the most accomplished directors of the section was its very first, Mary Acworth Evershed (1867–1949), who founded the section in 1930.

John and Mary Acworth Evershed with the staff of the Kodaikanal Observatory, c1907. Mary is third from right, second row from front, and John fourth from right. John Evershed was initially assistant to Michie Smith (third from left, second row), but became director in 1911. (Courtesy: Indian Institute of Astrophysics Archives, Bangalore)

Mary was born Mary Acworth Orr, the fifth of seven children, in Plymouth Hoe on 1 January 1867. Early in her life she spent five years in Australia – this led to her first published work, *An Easy Guide to Southern Stars*. Mary's major contribution to astronomical history was to produce a Historical Section Memoir, a directory of craters on the Moon. *Who's Who in the Moon* presented biographies of the many historical figures, some obscure, who had lunar craters named after them. She was also heavily involved in producing the history of the BAA's first fifty years.

Not just a historian, Mary was also an accomplished solar observer. Over her life she attended five total eclipses of the Sun, beginning with BAA expeditions to Norway in 1896 and Algiers in 1900. She had been offered a post at the Dunsink Observatory in Northern Ireland but instead in 1906 she married a fellow solar observer and eclipse chaser, John Evershed. Shortly afterwards, John became director of the Kodaikanal observatory in India, offering Mary an ideal opportunity to pursue her observational interests. She produced several papers on solar observations, some in conjunction with her husband; one, on sunspot prominences, in her sole name. With John Evershed she travelled to Kashmir and New Zealand to research astronomical sites. Her time as director of the Historical Section followed their return to England in 1923.

What I like about Mary Evershed was the range of her interests. At the age of 20 she visited Germany and Italy, and in Florence she began to study the works of Dante Alighieri. She was particularly interested in the astronomical allusions in Dante's Divine Comedy, which led eventually to a book *Dante and the Early Astronomers*. One of Dante's translators was Dorothy L. Sayers, with whom she was friends. In her later years Mary was renowned for her gardening skills in the home the Eversheds maintained in Surrey.

Mary Evershed was but one amongst many women who flourished in the BAA right from its foundation in 1890. The very first BAA council contained three notable women. Agnes Clerke (1842–1907), the astronomical historian, and Margaret Huggins (1848–1915), about whom more later, were general council members; Elizabeth Brown (1830–1899) was the first director of the Solar Section; she had been Solar Section director of the BAA's precursor, Liverpool Astronomical Society, and it was her efforts that led to the foundation of the BAA when Liverpool AS hit problems. Additionally, Annie Scott Dill Maunder (1868–1947) was mover and editor of the BAA's Journal from 1894–96, a post she held again between 1917 and 1930.

Annie Scott Dill Maunder was born Annie Russell in 1868 in Strabane, Northern Ireland. She had sat the Mathematical Tripos at Cambridge University but, as was the custom for women at the time, was not allowed a degree. In her professional life she was a computer, literally a human calculator, at the Greenwich Observatory – this was the highest-paid astronomical position that a woman could aspire to in the nineteenth century; in consequence the Greenwich computers were highly qualified and highly competent. Annie took daily measurements of sunspot numbers; she reported to Edward Walter Maunder, first assistant to the spectrographic and photographic department. Walter Maunder was arguably the first professional astrophysicist in Britain.

Annie Scott Dill Maunder. (Wikimedia Commons)

In 1895 Walter and Annie married, which meant that Annie had to give up her employment at the observatory. However, she continued to work on solar problems, and their marriage was a successful partnership professionally. The Maunders are renowned for two things – first, the famous 'butterfly diagram' which shows the latitude of sunspots varying predictably over an approximately eleven year solar cycle. This was a joint production of Annie and Walter. Secondly, the Maunder Minimum (the period between 1645–1715 marked by a dearth of sunspots – and very cold winters) was noted by Walter Maunder in a paper published under his name only, although as Annie had a longstanding professional interest in sunspots it's likely that she had some part in the discussions (Maunder also acknowledged earlier work by Spörer). Another joint production, which Walter Maunder acknowledged was largely Annie's work, was the 1908 popular astronomy book *The Heavens and Their Story*. Like the Eversheds, and Elizabeth Brown, and me, Annie and Walter Maunder were eclipse chasers; like Mary Evershed, Annie's first of five total eclipses was the BAA's expedition to Norway in 1896. In India in 1898 she photographed, 'the longest ray', the longest coronal streamer ever photographed to that date; her images of the solar corona in Mauritius in 1901 included solar plumes. The lunar crater Maunder is named for the two of them.

Mary Adela Blagg was another BAA founder member. She had two specialities, these being selenography (the study of the Moon) and variable star calculations. Mary was born in Cheadle, Staffordshire in May 1858. She received little formal education, as she had to bring up her siblings following their mother's early death; nevertheless, she taught herself mathematics to a high standard from her brothers' textbooks. In middle age, she became interested in astronomy and attended lectures in Cheadle by Joseph Hardcastle, grandson of Sir John Herschel. In 1906 she submitted a paper to the BAA Journal on the scintillation (twinkling) of stars; Hardcastle, recognising the quality and tenacity of the work which went into the observations, introduced her to Samuel Arthur Saunder, BAA President, and Prof Herbert Hall Turner of Oxford University. Turner had realised the necessity of standardising lunar nomenclature, and Saunder and Blagg began a long and fruitful collaboration, starting with *A Collated List of Lunar Formations* in 1913; by 1935 Blagg had extended this to an IAU publication *Named Lunar Formations*.

Mary Adela Blagg. (Courtesy of Jeremy Shears)

Another project suggested by Turner was the reduction and analysis of variable star observations. A lot of observations had been made during the late nineteenth century by astronomers such as Joseph Baxendell (a leading member of the Manchester Literary and Philosophical Society), but little had been done with them. Turner suggested that Blagg used Baxendell's observations to produce magnitude (brightness) estimates, and then analyse these harmonically to discover any periodicity in the light curves. Blagg turned out to be highly skilled at this time-consuming task. She and Turner produced a long series of papers in the *Monthly Notices of the Royal Astronomical Society* in which they analysed several stellar light curves, many from the observations of BAA Variable Star Section members. For example, Blagg was able to deduce that the orbital period of the eclipsing variable star Beta Lyrae was gradually increasing. She also applied harmonic analysis to Bode's Law, the numeric pseudo-law connecting the distances from the Sun to each of the planets, and produced 'Blagg's Formula', a generalisation of Bode's law which also encompassed satellite orbits in our solar system. (This paper appears in Volume 73, p414 of the *Monthly Notices of the Royal Astronomical Society*, immediately before Mary Evershed's paper on sunspot prominences on p422).

What is remarkable about Blagg was that she rarely ventured far from Staffordshire, preferring to do most of her work by correspondence. She was, however, persuaded to attend two IAU conferences, in Cambridge (1925) and Leiden (1928). Mary Evershed attended both these too. In her birth town of Cheadle, she is commemorated by an armillary sphere, built by James Plant and erected to mark the millennium in 2001. Mary Blagg also has a lunar crater named after her.

So, women were strongly involved in the British Astronomical Association right from the start. In this respect, the BAA stood in contrast to the Royal Astronomical Society, who, shamefully, did not admit women to fellowship until 1916. The first women to be admitted to fellowship were all BAA members – Mary Blagg, Fiametta Wilson (1864–1920), Alice Grace Cook (1887–1958), Ella K. Church and Irene Warner. Annie Maunder became an FRAS later in 1916, Mary Evershed in 1924 after her return from India. Wilson and Cook were joint-directors of the BAA's Meteor Section during the 1920s.

Admittedly there had been previous honorary fellowships given to women by the RAS. It would have been churlish of them not to admit Caroline Lucretia Herschel (1750–1848), one of the most successful ever comet discoverers, a winner of the RAS's gold medal, and an integral part of the Herschel family observing team who discovered so many nebulae during the late 18th and early 19th centuries. I am sure I need not point out that William Herschel was the prime observer of nebulae (although Caroline did discover several of her own), but his observing technique, involving long hours spent exclusively at the eyepiece to protect his dark adaptation, required skilled and equally tenacious technical backup. I ought also to mention a third member of the team, Johann Alexander Herschel (1745–1821), who was their telescope maker.

Mary Somerville (1780–1872), the polymath after whom Somerville College Oxford is named, was admitted to honorary fellowship of the RAS at the same time as Caroline Herschel. The scientific patron Anne Sheepshanks (1789–1876) followed in 1862. In 1903 two other women were admitted to honorary fellowship of the RAS, Agnes Clerke and Margaret Huggins. To my mind, both deserve to be far better known. Margaret Lindsay, Lady Huggins was a pioneer of spectroscopy. She was born Margaret Lindsay Murray in Dublin on 14 August 1848 although her mother died when she was young. Family lore has it that her father instilled a love of astronomy in her, and she also developed an interest in photography. Once again, her marriage enabled her to fulfil professional potential. She married William Huggins, twenty four years her senior, who had already made ground-breaking advances in spectroscopy; for example, measuring the radial velocities of stars by measuring the Doppler shifts of their spectral lines. William Huggins was in danger of being overtaken in the field by newer younger astronomers but in Margaret he found a collaborator to revitalise his career and to launch hers.

Margaret Lindsay Huggins. (en. wikipedia.org/wiki/Margaret_ Lindsay_Huggins)

William Huggins had previously observed spectra visually, making careful measurements by hand. Once Margaret began observing, spectra were recorded photographically. She also tried new and innovative photographic techniques such as using dry or gelatine plates rather than wet collodion. The Hugginses produced many joint papers, most notably *The Atlas of Representative Stellar Spectra* in 1899; their research interests included taking spectra of the solar corona, outside eclipse, and investigating the very distinctive 'single-line' spectra of nebulae. These were impossible to replicate in the laboratory, unlike so many lines in solar and stellar spectra, and were suspected to be associated with an as-yet undiscovered element, the so-called 'nebulium'. It turns out that these lines are actually due to 'forbidden' energy-level transitions in the extremely tenuous gases of nebulae. These investigations informed two of the great advances of twentieth-century physics: the realisation that some nebulae are gaseous (clouds collapsing into new stars, or the exploded remnants of old ones) and some are actually distant galaxies like our own. The mystery of why different elements had signature spectral lines was answered when atomic structure began to be probed early in the next century.

After William Huggins's death, Margaret fought tenaciously to promote the scientific legacy of her late husband; she was also a regular and respected attendee at the RAS. She donated the Huggins family archive and some instruments to Wellesley College, an all-ladies college in Massachusetts.

I hope that these short biographical notes have introduced you to some figures from the early days of the BAA and RAS who deserve to be much better known. It has been the recent job of historians of astronomy to highlight the careers of these pioneers.

References

Bob Marriott lectured on Mary Acworth Evershed in the British Astronomical Association Historical Section's meeting in November 2010. An account of his talk can be found in the *Journal of the British Astronomical Association*: JBAA Vol 121. No 2, pp. 110–112.

Mary Acworth Evershed's obituary appeared in the *Monthly Notices of the Royal Astronomical Society*: MNRAS 110, pp. 128–129.

A series of articles on the first lady members of the Royal Astronomical Society appeared in Astronomy & Geophysics in 2016, to commemorate the 100th anniversary of their admission to fellowship. Amongst them were:

> *Selenography and Variable Stars*, Jeremy Shears (A&G Vol 57, Part 5, pp. 17–18) is about Mary Adela Blagg.

> *A Pioneer of Solar Astronomy*, Silvia Dalla and Lyndsay Fletcher (A&G Vol 57, Part 5, pp. 21–23) is about Annie Maunder.

Edward Walter Maunder FRAS (1851–1928): his Life and Times, Tony Kinder JBAA Vol 118. No 1, 2008 pp. 21–42.

The Sun Kings by Stuart Clark (Princeton University Press, 2007) also has much to say about the Maunders.

Unravelling Starlight by Barbara Becker (Cambridge University Press, 2011), is a comprehensive biography of William and Margaret Huggins.

Is There Still a Place for Art in Astronomy?

David A. Hardy

It may be hard for today's astronomers to believe, but there was a time (and not so long ago, really) when all observations were recorded by *drawing* what could be seen through the telescope. Usually the medium used was a pencil, in various grades (HB, B, etc.), and these may have been later used as reference for painting or pen-and-ink. Even when photography came in there were still advantages to using this method, as photos needed long exposures during which the image would probably be disturbed by our turbulent atmosphere, leading to blurring, whereas the visual astronomer could take advantage of moments of clarity.

The astronomical telescope was first used around 1610. The best observers, like the Greek E.M. Antoniadi, were skilled artists in their own right. But until the end of the last century any attempt to depict a scene on another world was little more than fantasy, or at least allegorical in nature.

The illustrations by Emile Bayard and A. de Neuvill for the original edition of Jules Verne's *From the Earth to the Moon* (1865) were obviously science fiction, but were at the same time attempts to render the subject accurately according to the science of the day (which is all that any of us can do). The first space landscapes to appear in a non-fictional book were probably those by James Nasmyth, who in 1874 created both paintings and plaster-of-Paris models of the lunar surface, which were photographed against a black, starry background to illustrate his book with James Carpenter, *The Moon*.

By the turn of the century, periodicals such as *Cassell's Family Magazine*, *Pearson's Magazine* and *Pall Mall* were publishing illustrations by artists such as Stanley L. Wood, Fred T. Jane, Paul Hardy and Dudley Hardy (neither related to me, as far as I know). While science fictional, many of their illustrations are remarkably realistic.

A Frenchman, the Abbé Theophile Moreux, produced 'accurate' reconstructions of lunar scenes at about the same time, though his mountains often look more like church spires. He was Director of the Bourges Observatory, and wrote

James Nasmyth illustrated his book *The Moon* with James Carpenter (1874) with plaster models photographed against a black, starry sky.

and illustrated his own books, such as *A Day in the Moon* (1913). One of the most prolific and important astronomical artists of that period include the British illustrator Scriven Bolton (1883–1929), who followed Nasmyth's technique of photographing plaster landscapes, but added more detailed space backgrounds, such as the Earth in the lunar sky. He also painted and worked in pastels. He was born in Leeds and was a keen amateur astronomer. His full name was Thomas Simeon Scriven Bolton, and his untimely death came at the age of only 46. His work was seen in many publications, and notably in the *Illustrated London News*. As well as the Leeds Astronomical Society, at which he used their 26-inch reflector, he was a member of the British Astronomical Association and the Royal Astronomical Society. He was known to have met Bonestell in the 1920s, when the latter was living in London, and indeed Chesley probably started making and photographing model landscapes, then painting them, soon after, spurred by Bolton's technique.

Although he did also paint, British illustrator Scriven Bolton made plaster models very similar to those of Nasmyth. He worked for *The Illustrated London News*.

The American, Howard Russell Butler, was an excellent painter and his work is now in the collections of the Smithsonian Institution and the American Museum of Natural History.

Both Bolton and Moreux contributed to the magnificent two-volume *Splendour of the Heavens* (1923), along with other, lesser-known artists such as H. Seppings Wright. Bolton also worked on the *Illustrated London News* in the 1920s, where he was joined by another Frenchman, Lucien Rudaux, and later by the American Chesley Bonestell. Both influenced later generations of space artists. We'll come back to them later.

Other artists who were working in the 1930s and who produced many excellent examples of the genre include Rockwell Kent, who worked on *Life* magazine and in 1937 depicted four alternative 'end of the world' themes, Charles Bittinger, who illustrated a *National Geographic* project in 1939, and several basically science-fiction artists, notably Frank R. Paul and Charles Schneeman. SF artists have, in fact, often doubled as space artists, with varying degrees of success. (Bonestell's work often appeared on SF magazines, especially *The Magazine of Fantasy & Science Fiction (F&SF)*, but he always strongly denied being a science-fiction artist). Like Butler, Bittinger travelled with a solar eclipse expedition in order to record one of the few astronomical phenomena visible from Earth with the naked eye and produced spectacular results. More recently, US artist James Hervat emulated them.

Lucien Rudaux

The year in which Nasmyth and Carpenter's *The Moon* was published also happens to be the year in which Lucien Rudaux (1874–1947) was born. He began as an illustrator, but his hobby of astronomy led to his joining the Societe Astronomique de France at the age of 18. His observations of the planets appeared in its journal between 1892 and 1914. He founded an observatory at Donville, in Normandy, and produced a photographic map of our Milky Way galaxy. From 1900 magazines such as *Je Sais Tout* and *Illustration* published

'Mars from Deimos' by Lucien Rudaux, the first true astronomical artist, from his 1937 book *Sur les autres mondes*.

articles by him, illustrated with his own accurate impressions of landscapes of other planets. As an observer, he often concentrated on the edge or 'limb' of the Moon, so he knew that the lunar mountain ranges are rounded and eroded from micrometeorite impacts and extremes of temperature. This is the way he painted them in his landscapes, creating a sometimes startling resemblance to subsequent Apollo photographs but in stark contrast to Bonestell's craggy lunar landscapes.

Although Rudaux's work has the appearance of watercolour or gouache, astronomer/artist Dr William K.Hartmann has suggested that Rudaux might have used thin oils. He worked to a very small scale, and often in monochrome because his paintings would be printed this way as book illustrations. His style is quite impressionistic, but his detailed knowledge of the subject matter combined with his skill as an artist to produce truly realistic scenes. He took great care to calculate the relative size of a planet or the Sun in the sky, the shadow cast by the rings of Saturn upon the planet itself, and so forth. As well as working for the *Illustrated London News*, he wrote illustrated articles for the *American Weekly* supplement, covering such subjects as other solar systems, alien life and the end of the world (a subject which has fascinated a number of space artists, whose work also carries them back to the beginning of time).

Rudaux also produced a number of books, of which at least three are considered to be classics: *Le Manuel pratique d'astronomie* (1925), *L'Encyclopedie Larousse de l'astronomie* (1948 – continued after his death by Gerard de Vaucouleurs) and, most famous of all because it is the first book of space art, *Sur les autres mondes* (1937). The Larousse encyclopaedia was translated into several other languages, including English.

Lucien Rudaux was a council member of the Société Astronomique de France, and a member of both the Comité National d' Astronomie and the prestigious Union Astronomique International (generally known as the IAU). He was made a Knight of the Legion of Honour in recognition of his contribution to astronomy, and a crater on Mars has been named after him – perhaps the ultimate honour.

The Old Master of Space Art
Whether or not they have heard of Rudaux, people with the least interest in space art know the name of Chesley Bonestell, since he is without doubt the

Acknowledged as 'the true father of astronomical art', Chesley Bonestell (1888–1986) painted this Moon-landing in oils. It appeared in the 1949 *The Conquest of Space.*

most influential artist to date. Bonestell, pronounced 'bon-est-ell', (the name could even be translated as 'good star'!) was born on New Year's Day, 1888, in San Francisco, of a Spanish-Catholic family on his mother's side and an American-Unitarian family on his father's. (In later life Bonestell became an agnostic.). The Wright Brothers were then 17 and 21, H.G.Wells was 22 and not yet published, and Jules Verne was only halfway into his writing career. SF was still in its infancy. Yet Bonestell was able to see his visions of men walking on the Moon and probes visiting most of the major planets become reality, for he died on 11 June 1986. In 1985 an asteroid was named after him.

As mentioned above, it has come to light only relatively recently, largely thanks to artist and curator of the Bonestell estate Ron Miller, that Bonestell also made models, photographed them and painted over them in oils, creating amazingly realistic scenes of the Moon and planets. But his work became known for its jagged, craggy lunar landscapes, a result of his belief that without air, water or weather the mountains would be as rugged as the day they were born. He did not take into account the aeons of impacts by micrometeorites or the extremes of temperature, leading to the gradual flaking of rocks, which produced the rolling landscapes revealed by the Apollo landings in the 1960s and 1970s. Even so, his influence cannot be overestimated.

Bonestell's first painting of Saturn, done when he was 12, was destroyed in the San Francisco earthquake and fire of 1906. Rejecting his grandfather's wishes for him to become a businessman, he studied art in the evenings at Hopkins Art Institute, and in 1911, after studying at Columbia University, New York, became an architect. This stood him in good stead when, in 1922, he went to England and worked on the *Illustrated London News* and did advertising work for Lyons Restaurants. In 1938 he began work at RKO as a special effects or 'matte' artist, and worked on films such as *Citizen Kane* and *The Hunchback of Notre Dame*. But it was his work with producer George Pal on *Destination Moon*, *When Worlds Collide*, *War of the Worlds* and *Conquest of Space* that used his talents as a space artist.

His first published astronomical art was a series of paintings of Saturn from its (then nine) moons for a 1944 issue of *Life* magazine. Arthur C. Clarke was outraged by the comment of a shortsighted editor that 'the figures (of astronauts) are included only to give scale'!

He went on to lead a team of illustrators, working with scientist-writers under Wernher von Braun and Willy Ley, on a series of articles for *Collier's* magazine, showing how humans could explore space, from a wheel-shaped station in orbit to a fleet of moonships, and later a mission to Mars.

There can be no doubt that they succeeded spectacularly in showing a US public, whose lives had been dominated by Korea and the Cold War, a vision of a new frontier and great glory. In fact, they created a climate in which NASA could begin its work. Von Braun later designed the motors which launched America's first artificial satellite, Explorer 1, and the Saturn 5 which took Apollo astronauts to the Moon.

But Bonestell's highly realistic, even photographic oil paintings had another effect. As we have seen, his moonscapes showed dramatic, towering mountains – so much more inspiring than the flat, drab landscape on which Apollo 11 landed!

But could the disappointment of this reality have been a factor in the public's rapid disenchantment with the space programme? Throughout the 1950s, science fiction and other illustrators who had never even looked through a telescope produced variations of Bonestell's Moon (a warning to us all to do our own research and not copy the work of others…). Even so, the books which followed the *Life* and *Collier's* articles – *The Conquest of Space* (1949), *Across the Space Frontier* (1952) and *Man on the Moon* (1953; *Exploration of the Moon* in USA), among others, inspired future generations of space artists, including of course this writer.

Artists who combined SF and space art (because pulp magazines were the main market in the 1950s) include Mel Hunter, Alex Schomburg and Jack Coggins, all of whom produced excellent work. Schomburg, who died in 1998, was a master of the airbrush (which Bonestell did not use); his work is collected in *Chroma* (1986).

Other artists in the USA with their own individual style include Ken Fagg, Frank Tinsley – an aviation and science reporter who wrote and illustrated his own books on space travel, and Edward (Ed) I.Valigursky (1926–2009), who produced cover art for SF magazines, including a series of Solar System landscapes and other space subjects for *IF: Worlds of Science Fiction*. He produced his first space art in 1952 (as did I!). Neither should we forget the names of Dember, Pederson, McKenna, Emshwiller (Emsh), Ebel or Wenzel, all of whom made their contributions to the genre but whose full names were never given.

R.A. Smith

Ralph Andrew Smith (1905–59) is best known for his work with Arthur C. Clarke and the British Interplanetary Society (BIS) in the 1950s. He illustrated *Clarke's Interplanetary Flight* (1950), *The Exploration of Space* (1951) and *Exploration of the Moon* (1954) and his work can now be found in the BIS collection *High Road to the Moon* (1979).

Smith painted only to commission, never for his own satisfaction, and rarely produced anything other than space art. Like Bonestell, he had architectural training, but he was also an engineer and took great pains to ensure accuracy. He produced most of the designs for the lunar and ferry craft which are today associated with the BIS, who now own his work. He was thus a true 'first-generation' space artist – his work was all his own. His painting of a space station

'3000 Miles from the Moon'. Painted in 1955 by David A. Hardy. The moonship is a design by Ralph Andrew Smith of the BIS (© David A. Hardy / www.astroart.org)

in orbit was used in *The Perspectivist* magazine as an example of perspective drawing.

The first space artists had only telescopic observations on which to base their reconstructions. Later they were aided by the findings of astronomers using more sophisticated instruments, such as the spectroscope, which enabled them to know what elements are present in the light from celestial bodies.

Changing Landscapes

Mars was always seen as the most Earthlike planet. It has polar caps, and dark areas which appear to change with the seasons. Due to contrast with the reddish desert areas, these can appear greenish, so it was natural to assume that there could be vegetation, growing and spreading as the pole cap melts in spring. Some astronomers, notably Percival Lowell, even believed that they saw 'canals', dug by intelligent inhabitants to irrigate the deserts. . . Through telescopic observations using spectroscopes Mars was known to have an atmosphere; therefore it must have a blue sky – right? Likewise Saturn's huge moon Titan. Now the sky of Mars is orange-pink and glows down on craters and vast canyons instead of canals. Titan's sky is also orange, and at least partly opaque, with clouds, and an almost Earthlike landscape, but with seas and lakes not of water but of liquid hydrocarbons. Venus was long known to possess a dense, cloudy atmosphere which obscures its surface, so it could once be depicted with oceans of soda-water (because of the carbon dioxide in its atmosphere) or even with lush prehistoric jungles; or perhaps it was a wind-blown dust-desert, with rocks eroded into strange shapes? Now it is a blistering, hostile hell-planet, with massive shield volcanoes, sulphuric acid rain and lightning bolts.

It was believed that one side of Mercury always faced the Sun, so one side was bathed in the Stygian gloom of eternal night while the other roasted in the constant blaze of a huge Sun. Now it is cratered and remarkably Moonlike (Bonestell got this right), with a strange tide-locked rotation of 59 days and a year of 88. Jupiter used to have 11 satellites, while Saturn, with 9, was the only planet – perhaps in the entire Universe! – to be blessed with the unique phenomenon of rings. Today we know that all of the outer gas-giants possess such a halo, and their combined satellites number in the hundreds. But who could have forecast the active volcanoes of Io, the salty ocean below Europa's icy crust, the giant ice ravines on Miranda, or the geysers of Triton and Enceladus? Officially Pluto is no longer a planet but a 'dwarf planet', but this does not stop it from being one of the most interesting and perplexing bodies we have visited … Artists did not get it wrong; we could only work with the data that we were given! And we are delighted to find that the Solar System has turned out to be a much more exciting and dynamic place than was once believed.

About Art

A little digression to talk about art and illustration. To many people, one of the criteria for being an artist seems to be an ability to draw a straight line, or a perfect circle ("An artist – me? I can't even draw a straight line!" How often does one hear this said?) without mechanical aids. In the 19th century, John Ruskin certainly believed this, as his book *The Elements of Drawing* shows. But it surely depends upon the type of artist one wishes to be; an abstract painter whose work consists of broad strokes or squiggles of colour is unlikely to be concerned about such matters. On the other hand, as soon as artists began to use linear perspective in order to create the illusion of volume and 3D space within a 2D frame – which dates back to the early Renaissance (Filippo Di Ser Brunellesco, 1377–1446, is usually credited with its first use in art) – various 'aids' began to come into use to assist in finding vanishing points and so on. Artists used rulers, a compass and a set square. Devices like the *camera obscura* were also used to enable more accurate depictions, and it is known that Vermeer was fascinated by lenses and made use of them in his paintings, as did van Eyck. Of course, it was not until the invention of the telescope, first used astronomically

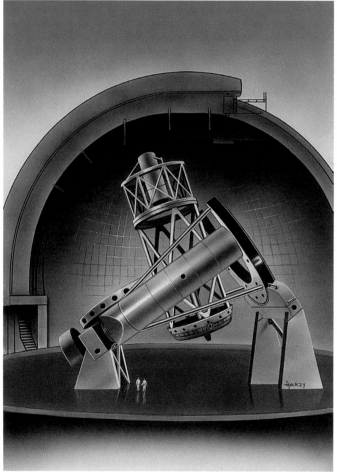

'Palomar Cutaway' by David A. Hardy. The 200-inch (508cm) Hale Telescope at Mt. Palomar, California, for 45 years the world's biggest telescope (© David A. Hardy / www. astroart.org)

by Galileo Galilei in 1609, that we had any concept of celestial bodies as other worlds, perhaps like the Earth.

The first use of photography in astronomy was as early as 1840, when Dr. Henry Draper succeeded in taking images of the Moon. By 1860, also in the USA, Dr. John Draper had obtained superior pictures of the Moon's craters with a 15½-inch refractor. Two English astronomers, Dancer and De La Rue, took some of the best lunar photos at that time.

In astronomy today, as in everyday life, digital photography has almost completely taken over from film. But a normal CCD (charge-coupled device) camera can generate unwanted 'noise' during long exposures, so for sharp images and colour accuracy 'cooled chips' are needed, together with a SLR (single-lens reflex) camera, all adding to their expense. A way around this is to take many short exposures and 'stack' them digitally, in a program like Photoshop, which can then also be used for cropping and editing. Today one can see results made in this way which far excel film photographs from ten or more years ago. *But* – it has to be said that the very best images still require some artistic input, for both their composition and colour (if any).

Of course, the images taken by probes and satellites provide detail and information that can never be obtained from Earth. Even so, the vast majority of these are taken *from* space, looking out or down at the surface of a planet or moon, or perhaps a comet or asteroid. Only a few bodies, such as the Moon, Mars, Venus and Titan, have yet been visited by landers or rovers which can record their landscapes from the surface, and few would deny that whenever possible a skilled human photographer will obtain better results than any robot! The only example of this is of course Apollo, but sadly the last photos taken there were in 1972 …

All of which means that actual scenes on other worlds, whether in our own Solar System or the 'exoplanets' of another star, will for a long time into the future require the expertise of artists (generally known as astronomical artists or space artists) who also have a good knowledge of astronomy and science. To get some idea of what these can do, visit the website of the International Association of Astronomical Artists (IAAA): **www.iaaa.org**. It was in June 1982, that nine artists from the USA and Canada first came together on the volcanic island of Hawaii and formed the basis for the IAAA. A year later, 19 artists met in Death Valley, California, and as the word spread, more artists,

'The Ocean of Space' by David A. Hardy. Our galaxy seen from the earthlike world of a binary star outside the main spiral (© David A. Hardy / www.astroart.org)

of both sexes, joined – not just from America but from Britain and many other countries. Today there are around 120 members. Not all are realists, and some produce work which is impressionistic, expressionistic, abstract or surreal, but the majority (unlike science fiction and fantasy artists, who work almost purely from imagination) do have a background in astronomy, physics and mathematics which enables them to interpret accurately the data from observatories and space probes, and convert them into believable scenes.

The latest and perhaps most exciting area for space artists is the discovery of extrasolar worlds or 'exoplanets'. In my 1972 book with Patrick Moore *Challenge of the Stars* I included paintings of hypothetical planets of the stars Proxima Centauri (our neighbour), Zeta Aurigae, Beta Lyrae and Zeta Cancri, a triple star system. But at that time these worlds were only surmised as a possibility, or at best a probability. It was not until the turn of the century (and millennium) that new, innovative and highly sensitive methods of observing began to provide us with proof that such worlds actually exist. The first bodies to be discovered, not surprisingly, were giants similar to Jupiter but often even larger. The search is now on for terrestrial or Earthlike planets. I was delighted to find that such a planet discovered orbiting Proxima Centauri B has a period of around 11 days, as the one mentioned above was postulated to go round its (red) star in 10 days, in order for it to have liquid water. Sadly the radiation in its environment does not make it a suitable candidate for life… The artist Lynette Cook has also been prolific in producing renderings of exoplanets, and this is undoubtedly a fertile field for future portrayals.

In a very real sense, space art has come of age. No longer stylized 'artists' impressions', the creations of today's artists are used to show how space will be developed; they present the surfaces of other worlds from angles impossible to achieve from space probes (and with greater clarity), and they depict objects which are at present 'real' only to theoretical astronomers, such as black holes and galactic cores. But their work also adorns the walls of connoisseurs and hangs in museums and art galleries. Space art is finding its own niche in the Art world, just as aviation, marine or western (cowboy) art have. (Oddly, there is perhaps more of a parallel with the latter, since although the actual landscapes often still exist, western artists have to use their imagination to reconstruct the past, rather than the future).

(Parts of this article have been adapted from *Visions of Space* by David A. Hardy (Paper Tiger, 1989).

Supermassive Black Holes

David M. Harland

Introduction

This article explains the discovery of supermassive black holes in the cores of galaxies, including our own, and the manner in which they evolved from initially voraciously swallowing up material to ultimately starving and entering a prolonged period of dormancy.

Quasars

On being hired by the Bell Telephone Company at Holmdel in New Jersey in 1928, Karl Jansky was assigned the task of identifying the sources of 'noise' that impaired the newly introduced transatlantic radio-telephone service. He built a large rotating antenna on nearby Crawford Hill, and set to work. Having eliminated all of the known sources, he realized in 1931 that there was a residual 'hiss' coming from the sky. By 1932 he knew that the source was strongest in the direction of the centre of the galaxy, which lies in the constellation of Sagittarius. Although Jansky published his results in 1933 in the *Proceedings of the Institute of Radio Engineers* which few (if any) astronomers consulted, the discovery was also featured in *The New York Times*.

In 1937 Grote Reber, a radio engineer in Wheaton, Illinois built a 9-metre diameter parabolic dish antenna on a tilting mount in his back yard to survey the sky. He started with the dish aimed at the zenith and then lowered it by a given amount each day, taking note of the varying signal strength as Earth rotated on its axis and scanned the dish across the sky. In 1938 he took the resulting map to the nearby Yerkes Observatory where Otto Struve, the director, noted that there was strong emission in the plane of the Milky Way. Furthermore, Struve realized that the dust that impeded the optical view was transparent to radio, and that Reber had detected sources lying *at* the galactic centre.

After the Second World War, 'radio astronomy' was pursued in England by A.C.B. Lovell at the University of Manchester, by J.S. Hey at the Royal Radar Establishment at Malvern, and by Martin Ryle at Cambridge University. In

A short-exposure of the giant elliptical galaxy M87 taken to show the prominent jet, whose blue light contrasts with the red starlight. It was taken using the 3-metre reflector at the Lick Observatory in California. In a longer exposure the jet would be swallowed by the outer regions of the galaxy.

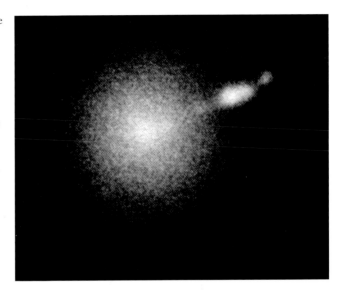

addition, it was pursued by J.G. Bolton, J.L. Pawsey and B.Y. Mills in Australia and by J.H. Oort in the Netherlands.

Hey inspected a strong radio source in 1945 which Reber had charted in Cygnus. On observing that its intensity fluctuated on a time-scale of minutes, Hey realized its source must be physically small. He named it Cygnus A.

Bolton was too far south to observe Cygnus A, but in 1947 he built an antenna atop Dover Heights, a cliff near Sydney, in order that a discrete source rising over the horizon would be able to be observed both directly in the sky and by its reflection off the sea. Applying the interferometry technique gave much finer angular resolution than the antenna could achieve alone. In this way, he associated the Virgo A radio source with M87, a giant elliptical galaxy which Heber Curtis of the Lick Observatory had found in 1918 to have a narrow luminous 'jet' emerging from its nuclear region.

In 1951 F.G. Smith in Cambridge pinned down the position of Cygnus A sufficiently for an optical search to be attempted. Smith airmailed the position to Walter Baade at the California Institute of Technology, who inspected it using the 5-metre Hale Telescope on Mount Palomar and found a rich cluster of galaxies, at the centre of which was a peculiar 18th magnitude object that appeared to have a distorted wispy structure and a double nucleus.

Having recently co-authored a paper speculating upon colliding galaxies, Baade concluded that this was such an event. A spectrum by R.L.B. Minkowski showed strong emission lines which implied the presence of a hot plasma. The fact that the radial velocities of the two components were identical ruled out the colliding galaxies idea. Although Cygnus A was one of the brightest radio sources in the sky, the 6% redshift meant it was almost 1 billion light-years distant. How could a plasma generate such a strong radio signal?

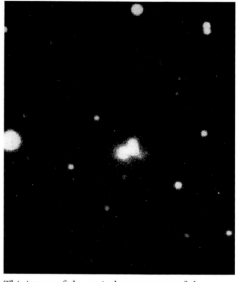

This image of the optical counterpart of the Cygnus A radio source was obtained by Walter Baade in 1951 by exposing a 'deep' plate on the 5-metre Hale Telescope on Mount Palomar. The double-lobe structure was unlike anything he had seen before.

In 1953 R.C. Jennison in England discovered that the discrete source was situated between two lobes of radio emission which formed a vast dumbbell shape. Whereas the discrete source was only 10 seconds of arc across, the lobes spanned 3 minutes of arc. Something in this galaxy was firing material 1 million light-years into space, and the process had been sustained for a considerable time.

Geoffrey Burbidge pointed out in 1958 that the radio emission indicated a non-thermal process such as synchrotron radiation by electrons in an intense magnetic field, which implied there was a 'powerhouse' of some sort in the core of the galaxy.[*]

Ryle's team in Cambridge used an interferometer to survey the sky and then published a series of catalogues. In their third catalogue, issued in 1958, each source was designated with the prefix '3C'.

---

[*] This is known as non-thermal radiation to distinguish it from thermal radiation emitted by a 'blackbody' at a specific temperature.

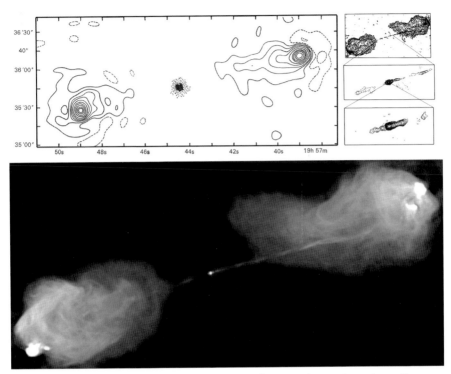

In 1953 R.C. Jennison discovered that the discrete source known as Cygnus A was situated between two vast lobes of radio emission. It was evident that something in this unusual galaxy was firing jets of material millions of light-years into space. As the spatial resolution of radio telescopes was improved, it became feasible to study the structure. The advent of computer processing made it possible to depict radio maps as visual images. A view at a wavelength of 6 centimetres by the Very Large Array in Socorro in New Mexico shows the narrow jets of relativistic electrons emerging from the compact core where the stars are located and the 'hot spots' made by the shock waves where the jets impinge upon the vast lobes of dispersing material. For further details see the paper 'Cygnus A', C.L. Carilli and P.D. Barthel, *Astronomy and Astrophysics Review*, vol. 7, pp. 1–54, 1996, or the volume *Cygnus A: Study of a Radio Galaxy*, C.L. Carilli and D.E. Harris (Eds), Cambridge University Press, 1996.

Whilst some radio sources correlated with galaxies, others were difficult to identify due to the substantial 'error box' on their radio position. However, in cases where the Moon travelled in front of a 'discrete' radio source it was possible to time when it was occulted by the Moon and, with results from multiple occultations, to refine the position sufficiently to search for the optical counterpart.

The 210-foot dish of the Parkes radio telescope viewed against the central section of the Milky Way. (Adapted from a CSIRO image taken by Wayne England)

Cyril Hazard monitored 3C273 in Virgo being occulted by the Moon in 1962 using the 64-metre dish of the Parkes telescope in New South Wales, Australia. This was only just possible, because the target was very close to the telescope's horizon and some trees had to be trimmed to produce a clear line of sight. The results not only refined the position of 3C273 sufficiently to permit an optical search but also revealed that it was a pair of sources separated by some 20 seconds of arc.

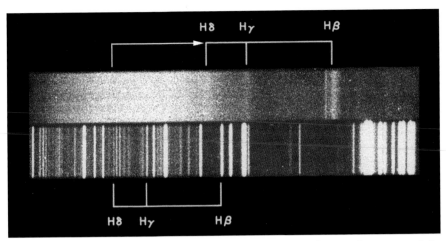

After recognizing the pattern of three hydrogen emission lines in the optical spectrum of the radio source 3C273 in Virgo (upper spectrum, with a comparison below) Maarten Schmidt at the California Institute of Technology was able to infer from their wavelength displacement a 16% redshift.

Allan Sandage at the California Institute of Technology made an optical search and found a 13th-magnitude star with a jet. One of the radio sources coincided with the star and the other with the tip of the jet. The situation resembled M87, but what kind of *star* could produce a jet?

Another Caltech astronomer, Maarten Schmidt took a spectrum of 3C273 in December 1962, which showed several spectral lines. In a moment of epiphany on 5 February 1963 he noticed the lines were correctly spaced for the hydrogen Balmer series, but with a 16% redshift. This implied a recession velocity of 47,000 km/sec which, if it was interpreted in terms of Edwin Hubble's redshift–distance relationship put the object *several billion* light-years off, deep in intergalactic space. If it was that far away, then its jet had to be enormously long. For 3C273 to appear as a 13th-magnitude star, it had to be 100 times more luminous than any of the known radio galaxies, the most powerful of which were the 'first ranked' members of their clusters at optical wavelengths. Schmidt wrote a one-page paper entitled '3C273: A Star-like Object with a Large Redshift' which was published in the journal *Nature*.

This became the first quasi-stellar radio source (soon shortened to 'quasar') and its discovery overturned the conventional wisdom about the origin and evolution of the universe.

One hundred and fifty quasars had been found by the end of the 1960s; and of the two-thirds whose spectra had been obtained, all were more heavily redshifted than 3C273, which turned out to be the nearest! Nevertheless, it was more remote than most galaxies known at that time. It was many years before improved telescopes, instruments and methods enabled us to glimpse the parent galaxies of some of the nearby quasars. Indeed, in some cases this required the upgraded Hubble Space Telescope.

The late 1960s saw the conception of the 'standard model' of the Big Bang. Although this predicted a 'cut off' in the population of galaxies at a high redshift, this transition was beyond the most distant measured galaxies and was therefore not evident, but the quasars offered a method of probing the early universe; possibly even to a time before galaxies formed. By 1973 only a few of the 200 redshifts were in the 80–90% category. This was consistent with a cut off, but the trend was ill-defined. In 1974 Schmidt and Richard Green began a search using the wide field of view of the 18-inch telescope on Mount Palomar, having equipped it with a filter to isolate intensely blue star-like objects. They discovered dozens more quasars that supported the case for a cut off.

When F.J. Low and Harold Johnson introduced an instrument that could detect objects whose peak visible light was redshifted into the infrared, these 'high redshift' quasars traced the overall distribution. There were none in our immediate neighbourhood, and only a few within several billion light-years. Farther out, the population progressively increased 1,000-fold, and then rapidly declined. This implied there was an evolutionary trend in which quasars developed early in the history of universe and were very common, but then evolved and fell dormant.

Black Holes

When Subrahmanyan Chandrasekhar found in 1930 that there was a maximum mass for a white dwarf star, he presumed that a larger star would suffer runaway collapse. Walter Baade and Fritz Zwicky speculated in 1934 that if a star which was too massive to become a white dwarf were to collapse, the electrons would be forced down into atomic nuclei, where they would combine with the protons to form a neutron gas. Like electrons, neutrons are subject to Pauli's Exclusion Principle. In 1939 J.R. Oppenheimer and G.M. Volkoff, one of his graduate students, calculated how 'degenerate neutron pressure' would resist further gravitational collapse and produce a 'neutron star'. Harlan Snyder, another of Oppenheimer's students, found that if this star exceeded about 3 solar masses, the pressure would be unable to prevent Chandrasekhar's runaway collapse.

Actually, this was not a new idea, because in 1796 the French mathematician P.S. Laplace had realized that if a collapsing cloud of interstellar gas was sufficiently massive it would form a *'corps obscur'* whose tremendous gravitational field would prevent light from escaping. Before the publication in 1916 of Albert Einstein's paper on general relativity, Karl Schwarzschild in Russia had been wondering about whether the universe could be described by non-Euclidean geometry. Whereas Einstein and Willem de Sitter had explored the cosmological implications, Schwarzschild mused on the prediction that the straight-line path of light would bend when passing close to a gravitating mass. He realized that if a star had a critical ratio of mass to radius, it would *warp* space so tightly that not even light would be able to escape. This became known as the 'Schwarzschild radius'.

This impression of the Cygnus X-1 system shows gas leaking from the blue supergiant star HDE 226868, whose shape is distorted by its companion's presence, and settling onto the accretion disk that circles a compact object of 5 to 8 solar masses that can only be a black hole. (Artwork by L. Cohen of the Griffith Observatory in Los Angeles)

The properties of such objects were later investigated in some detail but it was presumed that in practice the process of collapse would be disrupted by turbulence, preventing a 'singularity' from forming and instead giving rise to an irregular explosion. In 1965 Roger Penrose at the University of London proved that a runaway collapse *must* end in a singularity. In 1964 E.E. Salpeter at Cornell pointed out that the angular momentum of the material attracted by a singularity's gravity would cause it to spiral inwards to create an 'accretion disk' where the intense frictional heating would radiate X-rays.

In 1965 Y.B. Zel'dovich and O.H. Gusneyov in Russia reached this conclusion independently, and pointed out that if one star in a close binary system were to collapse to form a singularity, then the X-rays from the gas which it drew from its companion ought to be detectable.

At a New York conference in 1967, J.A. Wheeler of Princeton coined the term 'black hole' because the singularity constituted a veritable 'hole' in the fabric of space-time. The critical radius was labeled the 'event horizon' because no events occurring within it were visible.

When a sounding rocket was launched in 1965 with an X-ray detector, it observed a strong discrete source in Cygnus that was named Cygnus X-1. In 1970 the Uhuru satellite was launched into orbit for an all-sky survey at X-ray wavelengths. Cygnus X-1 was soon observed to be flickering on a millisecond time-scale which (by speed-of-light constraints) meant the source could be no more than 300 km across. Once radio telescopes had refined its position, optical attention focused on HDE 226868, a hot blue star of about 30 solar masses. Spectroscopic observations by C.T. Bolt of the University of Toronto showed that the X-ray source was orbiting this star with a 5.6-day period, and was 5 to 8 solar masses (this range being due to the uncertainty in the inclination of the orbit relative to our line of sight). The companion was too heavy for a condensed neutron star. It had to be a black hole. With this proof that stellar-mass black holes really existed, research took off.

Quasars and Radio Jets

In 1974, Martin Rees and Donald Lynden-Bell in Cambridge suggested quasars were accretion disks surrounding supermassive black holes in the cores of galaxies. Accretion disk processes can yield energy much more efficiently than nuclear fusion reactions. Nevertheless, quasars would have to convert between

0.02 and 20 solar masses per year to match the observed range of luminosities and (depending upon the rate, with the more voracious ones being most luminous and achieving the greatest masses) would grow to between 10^5 and 10^8 solar masses over 10 billion years of activity. In fact, in the most luminous examples the active region needed to be gravitationally bound by at least 10^8 solar masses to prevent the radiation pressure from fully dispersing the accretion disk.

Roger Blandford then pointed out that as plasma spirals down towards the black hole it will carry with it a magnetic field. This will 'connect' to the black hole and the field lines will 'wind up' very tightly until they momentarily 'snap', prior to reconnecting. During the time the field is 'open', plasma will travel along the magnetic axes, co-aligned with the rotational axis of the central black hole. This rapidly repeating cycle 'pumps' a succession of pulses of plasma into a magnetic 'tube'. This is how the jets are produced and sustained for periods lasting millions of years. The electrons in the plasma can be detected by radio because they are accelerated along the tube to a substantial fraction of the speed of light, making them 'hot' in the relativistic sense and hence able to emit non-thermal electromagnetic radiation.

The magnetic wall prevents the plasma from dispersing. A jet travels outward until it reaches a dense patch of the intergalactic medium, where the resulting shock wave heats that gas to create the 'hot spot' that is observed in the radio lobes. The shock front disperses material out to the sides to create the voluminous radio lobes.

If our line of sight views the galaxy more or less face-on, then we can observe the accretion disk as a quasar outshining its host. But if the galaxy is edge-on, the accretion disk is obscured from view and we perceive the radio lobes projecting to either side.

Galactic Cores

Since the total mass of a galaxy must far exceed that of any central supermassive black hole, the presence of such an object will make an insignificant contribution to the gravitational field that controls the motion of the material in the outer regions but it will dominate the nuclear region. In the absence of a supermassive black hole the orbits of the stars in the nucleus will be random (in contrast to the sedate procession of stars rotating in the plane of a spiral) with the stars swarming about like bees. If a black hole is present, however, it will draw material in and the angular momentum of that material will cause

The large spiral galaxy in Andromeda, M31 (NGC 224), with its two small companions M32 (NCG 221; lower centre) and M110 (NGC 205; upper right). The image was taken in 1979 by Vanessa Harvey using the Burrell Schmidt Telescope on Kitt Peak operated by the Warner and Swasey Observatory of Case Western Reserve University. (Image courtesy of Bill Schoening, Vanessa Harvey / Research Experience for Undergraduates Program and NOAO / AURA / NSF)

it to form a rapidly rotating disk. A disk of either gas or stars at the dynamical centre of a galaxy is therefore a tell-tale sign of a supermassive black hole, since without that intense gravity the disk would disperse.

In principle, the rotation profile of the material in the nuclear region of a galaxy can be measured by orienting a spectrograph's slit across it to determine the distribution of radial velocities, but few ground-based telescopes have

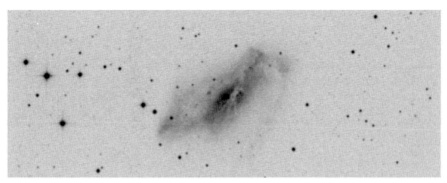

The optical counterpart of the radio source 3C120 as documented by the Palomar Observatory Sky Survey. Note its extremely bright (star-like) nucleus (dark in this negative image) and the irregular wispy peripheral structure.

sufficient resolution to isolate even the nuclear region of the M31 spiral galaxy in Andromeda, which spans 1 second of arc.

Nevertheless, in 1983 Alan Dressler at Caltech decided to look for supermassive black holes.

In 1943, C.K. Seyfert issued a catalogue of a dozen otherwise normal spirals which possessed unusually bright nuclei. The most prominent was M77, also listed as NGC 1068 and associated with the Cetus A radio source. Others were discovered later, some of which were radio sources. One, 3C120 in Camelopardalis, bore a striking resemblance to a quasar but its redshift of only 3% gave rise to speculation that Seyferts (as they became known) might be the 'missing link' between quasars and normal galaxies. Such a relationship was strengthened by the fact that the luminosity of a Seyfert nucleus can vary on a time-scale of several weeks. If one were to be viewed from afar then (it was argued) only its nucleus would be visible and it would be classified as a quasar. Seyferts were subsequently redefined as a subtype of a broader group of 'active nuclei' galaxies, most of them ellipticals possessing jets.

M77 was a strong candidate for having a central black hole, so Dressler selected it to be his first target. For comparison purposes he decided to inspect M31, which there was no reason to believe had one.

M77 proved too far away for the Hale Telescope to resolve its nuclear region, but Dressler was astonished to find a disk of stars deep in M31 circulating at 150

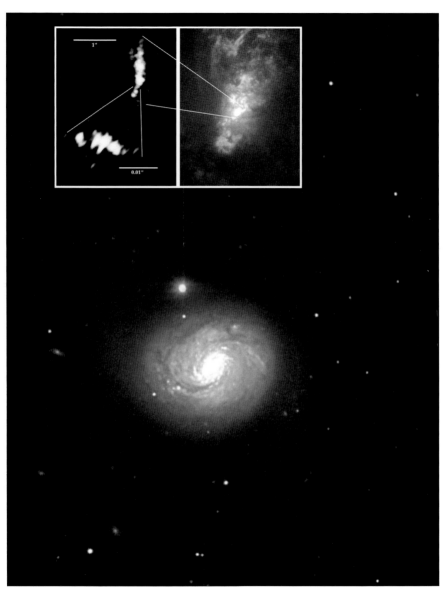

The radio source Cetus A is associated with the bright nucleus of the Seyfert galaxy M77 (NGC 1068). The Hubble Space Telescope resolved the nuclear region (insert, right). Polarisation implied the presence of an obscuring torus of material and this was verified by very long baseline interferometer of the radio jets (insert left). (Main picture courtesy of Michael Chase, Adam Block/ NOAO / AURA / NSF)

km / sec, implying the presence of a black hole of about 30 million solar masses. A supermassive black hole in the heart of a galaxy which showed no sign of *ever* having been 'active' was a major revelation.

A study by John Kormendy of the University of Texas at Austin on the Canada-France-Hawaii Telescope on Mauna Kea in Hawaii not only verified M31 but also identified a number of others black holes, including one of 3 million solar masses in M31's diminutive satellite M32 and another of 1 *billion* solar masses in NGC 3115, a lenticular galaxy in Sextans nicknamed the Spindle because it is an edge-on spiral.

In 1994, after astronauts installed a corrective optics package to remedy the spherical aberration of its primary mirror, the resolving power of the Hubble Space Telescope revealed supermassive black holes in the cores of more distant galaxies. The rate of discoveries increased after an imaging spectrograph was added to the Hubble Space Telescope in early 1997, with black holes being identified in all sorts of galaxies.

The various morphologies of galaxies derive from dynamical considerations. A galaxy such as our own, or M31, comprises an almost spherical nuclear region (named the 'bulge') whose centre is coincident with a broad disk of gas and

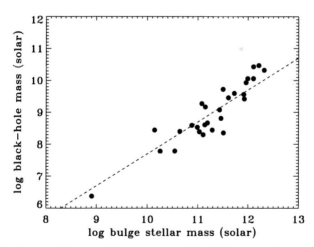

The relationship between the mass of a supermassive black hole in a galactic nucleus and the surrounding bulge, the mass of the latter being derived from luminosity measurements by the Hubble Space Telescope. (Adapted from John Magorrian *et al.*, 'The Demography of Massive Dark Objects in Galaxy Centers'. *Astronomical Journal*, vol. 115, pp. 2285–2305, 1998)

dust, and the stars that coalesced from it. An elliptical galaxy has no disk, it is 'all bulge'. This is not to say that ellipticals are smaller than spirals, because giant ellipticals, such as M87, are *larger* than most spirals.

As more results were obtained, it was discovered that the larger galaxies had more massive black holes. In fact, there was a linear relationship between the mass of the black hole and the luminosity (and therefore the mass) of the bulge of the host galaxy.

Interestingly, a relationship of this sort had just been predicted by Joseph Silk and Martin Rees. The conventional view was that the primordial gas created by the Big Bang had simply condensed to create stars and galaxies, and the great debate had been whether this process had occurred in a top-down (galaxies first) or bottom-up (stars first) manner. However, Silk and Rees proposed that the centre of each primordial gas cloud collapsed to form a black hole which, as it fed on the nearby gas, 'switched on' as a quasar, and the radiation pressure

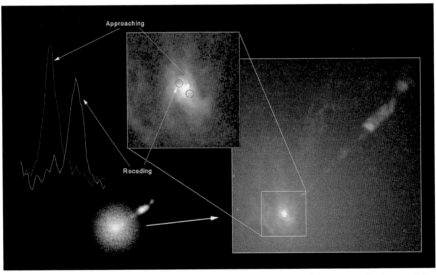

The Hubble Space Telescope found a disk of gas 500 light-years in diameter at the heart of the giant elliptical M87 which had a radial velocity profile implying the presence of a central black hole with the incredible 3 billion solar masses. (Adapted from a press release by Holland Ford of STScI/Johns Hopkins University; Zlatan Tsvetanov, Arthur Davidsen, and Gerard Kriss of Johns Hopkins University; Ralph Bohlin and George Hartig of STScI; Richard Harms, Linda Dressel and Ajay K. Kochhar of Applied Research Corporation; Bruce Margon of University of Washington in Seattle; and NASA/ESA)

then induced intense shock waves which in turn stimulated prodigious star formation. The process of galaxy formation was therefore a direct consequence of the black hole's presence. The quasar would 'switch off' when the black hole was 'starved', which Silk and Rees said would occur when the quasar became so luminous that its radiation pressure literally drove the rest of the material beyond the black hole's gravitational reach.

As the survey was extended, it was found that giant ellipticals such as M87 had central black holes of 3.3×10^9 solar masses; a large spiral such as M31 had one of 3.5×10^7 solar masses, which was only slightly greater than that of the Seyfert M77.

The Centre of the Milky Way

The centre of the Milky Way is hidden by the optical extinction of intervening gas and dust. The first indication of its nature was the discovery by Jansky that it was a strong source of radio emission. In his follow-up observations a decade later, Reber designated it Sagittarius A. For many years, the centre of the galaxy remained the province of radio astronomers.

In 1974 Bruce Balick and Bob Brown used the National Radio Astronomy Observatory's interferometer at Green Bank in West Virginia and discovered a strong point-like synchrotron source in the heart of Sagittarius A whose unusual characteristics made its nature a puzzle. They named this Sagittarius A* (abbreviated to Sgr A*, and pronounced 'A-star').

Meanwhile, Robert Haymes of Rice University in Texas flew a gamma-ray detector on a high-altitude balloon and noted a source in the general direction of the galactic centre. Its location was not easy to pin-point, because early gamma-ray detectors didn't have very good angular resolution. Nevertheless, the energy spectrum indicated the annihilation of electrons by antielectrons. A balloon could collect data only for a short period. Sustained observations by the HEAO-3 satellite in the 1980s revealed the source to vary on a time-scale that meant its diameter could not exceed several light-months. The gamma rays were fully consistent with Sgr A* being a black hole.

Shortly after the gamma rays were first detected, the galactic centre was evident on the Uhuru satellite's all-sky X-ray survey, but it was not until the launch of HEAO-2 in 1978 with an X-ray imaging system that it became possible to localize the source to within 1 minute of arc of the galactic centre.

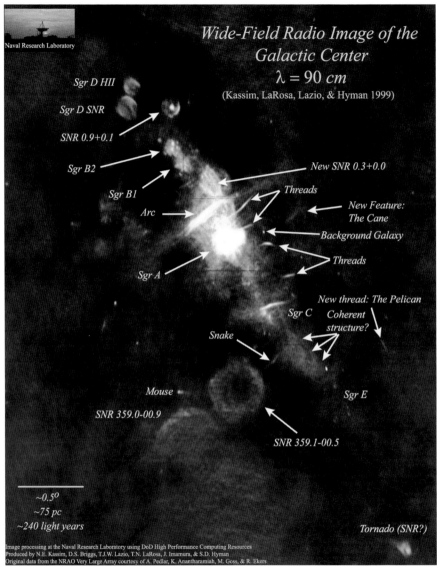

This is the largest and most sensitive radio image ever made of the central region of the Milky Way in the constellation of Sagittarius at a uniform and high resolution. It was compiled by N.E. Kassim at the Naval Research from Very Large Array observations at a wavelength of 1 metre. The galactic plane runs from lower right to upper left and passes through Sgr A. The bubbles are supernova remnants. The strong radio source Sgr A* embedded deep within Sgr A is precisely at the dynamical centre.

The Very Large Array is a radio interferometer consisting of 27 dish antennas, each 25 metres wide, configured with the 'arms' extending to the north, southeast, and southwest. The dishes travel on rails and the overall span of the array can be varied between 1 and 36 km as required for an observation. (Image courtesy of NRAO/AUI/NSF)

In this region of the spectrum, the variability implied a source approximately 3 light years across.

The Very Large Array in Socorro, New Mexico first turned its array of 27 radio dishes towards the centre of our galaxy in 1981. Exploiting its high angular resolution, in 1983 a team led by Ron Ekers saw for the first time a spiral of hot gas centered on Sgr A★. That same year, K.Y. Lo and Mark Claussen made an even more detailed map and refined the position of Sgr A★. In another study, Don Backer and Dick Sramek repeatedly measured its position in the sky in order to set a limit on its mass. If it was comparable in mass to a star, then it would be seen to travel rapidly in an orbit around the centre of the galaxy, but would travel more slowly if it was very massive. After 16 years of observing it to be *motionless*, they concluded that it was located right *at* the dynamical centre, a distance of some 25,000 light-years from us.

In 1984 Farhad Yusef-Zadeh, Mark Morris and Donald Chance discovered parallel filaments and threads, evidently following the magnetic field in that region. Far-infrared observations indicated ionized gas in a torus some 10 light-years wide. In 1998 the Very Long Baseline Array (which used 10 sites on the North American continent to synthesize an aperture thousands of kilometres wide and thereby achieve a very high angular resolution) found Sgr A* to be an elongated structure several hundred million kilometres in diameter.

In making pioneering observations in the near-infrared in 1968 Gerry Neugebauer had discovered a star cluster (IRS 16) that was coincident with Sgr A*. In 1986 an improved detector resolved it into red giants packed together at the equivalent of 1 million stars per cubic light-year.

The most direct way to determine whether Sgr A* was a black hole was to study the motions of the nearby streamers of ionized gas. While the radial velocities implied that 6×10^6 solar masses were concentrated within 10 seconds of arc of the Sgr A* radio source, the case was inconclusive because this was quite a large volume and the gravitating mass might be the cluster which was located within a few light-years of the centre. The only way to be certain was to measure the motions of the stars within IRS 16 by high-spatial-resolution infrared astrometry.

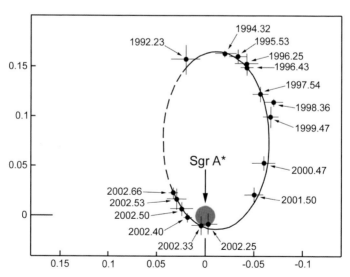

The 15-year eccentric orbit of the star closest to Sgr A* indicates a black hole of 3.7×10^6 solar masses at the dynamical centre of our galaxy. The axes centred on Sgr A* are in seconds of arc, and the periapsis in 2002 was at a range of 17 light-hours, equivalent to a mere 124 AU. (Adapted from 'A Star in a 15.2-year Orbit around the Supermassive Black Hole at the centre of the Milky Way' by R. Schödel *et al. Nature*, vol. 419, pp. 694–696, 17 October 2002)

In the early 1990s, two teams began using 'speckle' imaging to track the proper motions of the stars deep within IRS 16: one team, headed by Reinhard Genzel of the University of Munich, Germany, used the 3.6-metre New Technology Telescope at La Silla in Chile, which had been inaugurated by ESO only a few years earlier; and the other team, led by Andrea Ghez of the University of California at Los Angeles, used the 10-metre Keck Telescope on Mauna Kea in Hawaii. The fact that stars located so remote from us displayed significant proper motions confirmed the presence of a black hole in the heart of the cluster. How large was it?

As the observations accumulated over the years, one particular star traced a large portion of a 15-year elliptical orbit and analysis of the periapsis passage in 2002 implied a gravitating mass of 3.7×10^6 solar masses which, in such a confined volume, could *only* be a black hole. The key properties of a black hole are its mass, its rate of rotation and its electrical charge. The radius of an event horizon depends upon the mass of the black hole. In this case, it can be no more than 15 million km in diameter. The radio source seems to be ionized gas in the inner region of the accretion disk. The elongated shape implies the disk is being viewed obliquely.

Although the black hole has consumed everything within easy reach and is now 'starving', the Chandra X-ray Observatory has observed occasional flickering which indicates ongoing irregular infall of material to the accretion disk. Fortunately for life in the galaxy, quasar activity is unlikely ever to resume.

In Conclusion

Having discovered supermassive black holes to be common in the centres of galaxies, astronomers were delighted to find a dormant one in our own galaxy that they can study in detail right across the spectrum.

One audacious observation, scheduled for 2017, is to operate as many antennas as possible across the globe as a very long baseline interferometer network using the most sensitive detectors to study Sgr A* at a wavelength calculated to penetrate the dense material in the accretion disk in order to directly resolve the event horizon of the black hole. Even for a planet-sized interferometer dubbed the Event Horizon Telescope, it will be an extremely challenging task to resolve an object at the galactic centre that spans just 10 micro-seconds of arc.

Modern Video Astronomy

Steve Wainwright

What is Video and Video Astronomy?

Video Astronomy is the use of video techniques to produce live images of astronomical objects, even deep sky objects, on monitor screens for real-time viewing. It is also the use of video techniques to capture high quality images of these same astronomical objects.

What is Video?

Video dates back to the 1940s and the beginnings of television. In those days it was a purely analogue system that has since evolved over time. Analogue video components are still used at the time of writing, but the whole system has evolved into the digital age; moreover, many of the features of digital video have their roots in the analogue systems of the past. For example the YUV colour format was introduced for analogue colour television, but has persisted in modern digital imaging systems. In its simplest terms, video is a stream of sequential image data produced by a camera and ending up in a range of possible destinations such as a video recording system, a computer display, a video monitor, or a television set, after the data have been transmitted wirelessly, by cable or via the Internet from the source camera.

It is difficult to know precisely when amateur video astronomy began, the best estimate that we can make is when it started to be used by a significant number of amateur astronomers. This was in the mid to late 1990s. At this time, astrophotography was done mainly by SLR (Single Lens Reflex) film photography and a small number of CCD (Charge Couple Device) cameras. The CCD was invented in 1969 in the AT&T Bell labs. By 1971, the first images were captured by simple linear devices and by 1974 the first 2D array CCDs had been produced. These devices were the forerunners of the imaging chips (sensors) used today. The first still CCD camera was made in 1975 using a 100 x 100 pixel array, a miniscule 10 kilopixels, compared with today's multi megapixel sensors. In 2009, 40 years after its invention, Willard S. Boyle and George E. Smith were awarded the Nobel Prize for inventing the CCD.

These days, multi-megapixel, sensors are used by amateurs. In 1989, the French Astronomer Christian Buil published a book entitled *Astronomie CCD, Construction et Utilisation des Caméras CCD en Astronomie Amateur*. An English edition was published in 1991. In those days it was possible to buy CCD cameras constructed for astronomy, but the cost could be many thousands of dollars; even a CCD chip alone could cost hundreds of dollars. His book shows the construction and operation of CCD cameras, as well as the theoretical and practical application of image processing. However, early in the book, Christian rejects the use of camcorders (video cameras) because of their fast frame rates and their bit depths (most were 8 bit devices). Whilst these were reasonable comments to be made in the early '90s, the evolution of the technology and software has rendered these comments obsolete. The production of CCD assembly lines and surface-mount technology resulted in a dramatic fall in the cost of CCDs and the devices that would contain them. There was also a competing sensor technology called CMOS (Complementary Metal Oxide Semiconductor) invented by Eric Fossum and co-workers; and in the mid '90s, NASA's Jet propulsion laboratory produced a number of prototypes. To begin with, CMOS sensors were less sensitive and more noisy than CCD sensors. However, the CMOS technology has matured and now rivals and even surpasses CCD technology. For most applications, CMOS sensors have now replaced CCD sensors. CMOS is more easily scaleable, is an easier manufacturing process and requires less support circuitry. Sony stopped the manufacture of CCD chips at the end of March 2017, but did guarantee deliveries of certain high end chips into the 2020s. Just as CCD sensors replaced the Vidicon cathode ray tube sensors of the early video cameras, so at the time of writing, CMOS sensors have effectively replaced CCD sensors.

Cameras used for Video Astronomy

8 Bit Devices

The earliest cameras used for video astronomy were camcorders, surveillance cameras and web-cams, although having said that, all of these devices are still used at the time of writing. What has changed over time is the design of the devices and the software that we use to exploit them. There are various configurations by which a camera can be connected to a telescope. The simplest configuration is the afocal configuration, i.e. through the eyepiece. The camera still has its lens in place and it is held to the eyepiece and the camera records the light that comes through the eyepiece. It is possible to hand hold a camera for an afocal shot, but there are available, a number of afocal mounts designed to hold the camera firmly in position at the eyepiece, such as the digiscoping mounts used by naturalists to photograph wildlife through spotting scopes.

Camcorders and compact digital cameras are typically used afocally as their lenses cannot be removed. DSLR (Digital Single Lens Reflex) cameras can have their lens removed so that the camera can be placed directly at a focus of the telescope, such as the Newtonian focus of a Newtonian reflector or the Cassegrain focus of a Maksutov telescope. DSLR cameras are included here with 8 bit devices even though they usually have 12 bit or 14 bit ADCs and are capable of producing RAW image data that can be converted to 16 bit TIFF files. However, they also save 8 bit JPG images. Some DSLRs can be computer controlled and their output can be viewed on a computer screen. Camcorders, Compact and DSLR cameras can be used, not only for the capture of still images, but also in movie mode for the capture of movie files. Having a built in viewing screen, they can be used for the live viewing of images of the Moon and planets. The movie format will usually have to be converted into a format suitable for stacking. For public viewing, and 'sharing the eyepiece' the small viewing screens of these devices are suitable for a small number of people.

A dashboard camera with the lens removed, and a webcam adapter can also be used for sharing the live viewing on its built in screen. Some of these cameras have additional AV output sockets and so can also display the image on a separate monitor. These devices vary considerably in price and suitability to be adapted for telescope use. They can give pleasing live views but the quality of any still images that may be produced will depend on the degree of compression imposed by the camera on the video feed. Devices such as this can be very useful when observing with a disabled person who may not be able to get in a position to look directly through an eyepiece.

Web-cams were some of the earliest low cost devices used to produce live views of solar system objects. The early devices to be used were cameras such as the ToUCam and later the SPC900NC. These were very good CCD cameras and were also suitable for the Steve Chambers, SC long exposure modifications that introduced long exposure astrophotography at a low price. All you needed was a steady hand, a handful of cheap, off the shelf components and a soldering

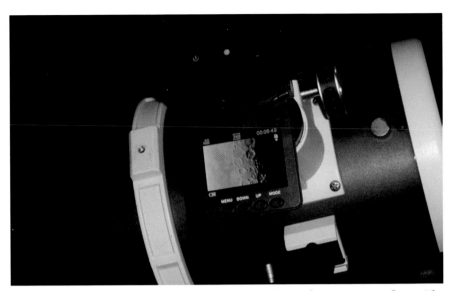

Plate 1: Dashboard camera mounted at the Newtonian focus of a Newtonian reflector. The pleasing image of the Moon displayed on the screen of the dashboard camera has a pink tinge which is due to the fact that the built in IR cut filter was removed with the lens. This upsets the white balance of the image which can be rectified if required by fitting an IR/UV cut filter to the Mogg scope adapter. The scope itself can be rotated in its rings to point the screen in the appropriate direction.

iron. With these cameras it was possible to see 'live' images of deep sky objects on your computer screen, where 'live' means the displayed image was updated every time a new image came in and was suitable for outreach and 'eyepiece' sharing.

In the early-2000s companies in the USA such as SAC Imaging, and in Europe, Perseu, did the long exposure modifications and offered the long-exposure modified webcams remounted in purpose-built housings. This work and much more stemmed from the international Imaging group QCUIAG (**www.qcuiag. org.uk**) that I set up in 1998 and which, at the time of writing, has more than 9,000 members worldwide. Unfortunately, most of these excellent CCD web-cameras no longer have driver support under Windows. However, they are still usable under Linux, using software such as our AstroDMx Capture for Linux. All modern webcams are CMOS based cameras, and many of them are totally unsuitable for astronomical use. In an attempt to produce fast frame rates, which are important for video conferencing, the intended use of webcams; the video stream is often highly compressed. Although this compressed video stream gives acceptable live views of the Moon and planets; if captured video frames are stacked in software such as Registax or Autostakkert!, the resulting image has compression artefacts that make it unusable. There are some webcams that do produce video streams with little or no compression. One of these is the Sweex WC066 that delivers video with very little compression but is very temperamental when initialising on the USB bus. It is handy for capture software to produce live streaming that does flat field correction before the video stream is displayed and/or captured. AstroDMx Capture for Linux can do this as well as on-the-fly dark frame correction. It is a pleasure to see 'dust doughnuts', vignetting and hot pixels disappear as the on-the-fly corrections are applied.

Surveillance cameras can be used for live observing and image capture via USB capture cards. These cameras produce an analogue video stream, usually in PAL or NTSC format; PAL is the superior format. These cameras can be plugged directly into a video monitor, usually with composite video input. If required, such a camera can be attached to a computer by means of a USB capture card and then effectively becomes a USB camera. There are other formats of surveillance camera that directly produce a USB output and require a computer to render the video stream on the computer display. However, some of these cameras apply some degree of compression to the output, so care needs to be taken when purchasing.

For several years there has been a newer type of analogue video surveillance camera that does frame accumulation and even genuine long exposures on the camera whilst outputting a continuous video stream at 25 frames per second for a PAL device; produced as interlaced half-frames at 50 half-frames per second. These cameras produce exposures or integrations that are exponentially increasing or decreasing multiples or fractions of frame times and the exposure times are normally expressed in this way: x1, x2, x4, x8, x16, x32, x64, x128, x256, x512 etc. or x1, x1/2, x1/4, x1/8, x1/16, x1/32 etc. The frame accumulation works by accumulating the frames into an internal camera buffer. As more frames are added to the buffer, the image brightens until the deep sky object to be observed or imaged, appears clearly on the display. The display updates at a rate depending on the number of frames being accumulated for long exposures and at 1/25s for short exposures. Such frame accumulating video cameras are very good for outreach because they do not have to be connected to a computer if that is not required. They can be connected directly to a video monitor or TV that has composite video input. They can also be connected to a DVD recorder and the session recorded in high quality for later playback or for extracting images from the DVD VOB (Video object) files, for stacking and image production. VOB extractor, written by Ian Davies, is a program that extracts BMP images from VOB files. It can extract unique frames produced by frame accumulating video cameras as the video stream is updated depending on the integration level. Examples of this type of video camera are Mintrons, Gstar-EX, Malincam, Samsung SDC-345 and SCB-2000, Watec, the LN300 and its derivatives.

Plate 2: The Swan nebula (Messier 17) imaged through a 6 inch, f/5 Newtonian using an LN300, frame accumulating surveillance camera. Viewed on a laptop screen. The camera was connected to the computer via a capture card and a pleasing, live view of the nebula could be seen whilst capturing frames for later stacking into a good quality image.

The More Bits the Better

The cameras available for video astronomy have different ADCs. That is, they deliver images with different bit depths. Predominantly, the ADC will deliver 8 bits, 12 bits, 14 or 16 bits output. An 8 bit image can contain up to 2^8 levels of brightness per colour channel, i.e. 256 levels, ranging from 0 (black) to 255 (white). A 12 bit image can contain up to 2^{12} levels of brightness, i.e. 4,096 levels, ranging from 0 (black) to 4,095 (white). Similarly, a 16 bit image can contain up to 2^{16} levels of brightness, i.e. 65,536 levels, ranging from 0 (black) to 65,535 (white). For a camera with a 12 bit ADC there are no readily available 12 bit image formats, so the 12 bits of depth either have to be placed in a 16 bit image container, such as a 16 bit Tiff, or it can be (wastefully) scaled down into an 8 bit image container such as an 8 bit Tiff. There is often confusion as to what the benefits are of having a 16 bit image when it will be typically displayed on an 8 bit monitor. The benefits are not so much in the display, but more in the type of processing that can be done and the effects of that processing.

Most deep sky objects have both brighter and much fainter regions. Processing of images of these objects aims to stretch the dimmer parts to make them brighter, but at the same time, not stretching the brighter parts to make them saturated. The greater the bit depth of an image, the more readily can the fainter parts of the image be stretched without causing artefacts.

Plate 3: 8 bit gradient.

Plate 4: 16 bit gradient.

Consider two images of gradients that appear to be identical, Plates 3 and 4. Both gradients range from black to white. However, one image is 8 bits in depth and the other is 16 bits in depth.

The two images in Plates 3 and 4 appear to be identical. However, if they are taken into an image manipulation program such as Photoshop or the Gimp, and the middle 10% of the gradient is stretched from black to white, Plates 5 and 6, the difference between the images becomes apparent.

Plate 5: Middle 10% of the 8 bit gradient stretched.

Plate 6: Middle 10% of the 16 bit gradient stretched.

The value of using 16 bit images when stretching the levels of brightness is apparent. The middle 10% of the 8 bit gradient produced pronounced banding when stretched, whereas the middle 10% of the 16 bit gradient is still quite smooth after stretching.

Summing 8 bit images into bit-deep image containers to obtain a bit depth advantage and simultaneously reduce noise

Around the turn of the millennium, a German amateur astronomer, Juergen Liesmann produced a Java program to capture frames from a low light surveillance camera and sum them into 32 bit deep FITs files. In a PAL video system the frames were being captured at 50 half-frames per second. The software worked fine, but was too resource hungry for the average laptop of the day. A 32 bit deep Fits file was chosen because it would be possible to sum more than 16.8 million 8 bit video frames into the file without saturating the image. In practice, it would be impossible to fill such a file and cause image saturation unless the individual images were saturated. Juergen used to sum many thousands of frames into the 32 bit Fits files. The resulting images could be scaled and stretched and they revealed the deep sky objects he was imaging, effectively with very short exposures, but using many short exposures to synthesise long exposures. Inspired by this work, Bev Ewen-Smith at the COAA observatory in the Algarve wrote a Windows program to do a similar thing. This software is called AstroVideo and goes beyond Juergen's work in a number of ways. Time delays in the capturing of video frames allow unique frames to be captured from frame integrating video cameras whilst a stack and track feature allows the resulting images to be co-registered while being summed. Post capture stacking is also built into the program. This is feature-rich software and still runs under Windows 10.

In our own Linux capture software, AstroDMx Capture for Linux, coded by Nicola Mackin; summing of 8 bit true long exposures into 16 bit Tiff or SER files allows the synthesis of hybrid long exposures by summing (off-chip integrating) 8 bit exposures of several seconds into 16 bit containers, to produce images of greater bit depth, that can be stretched a little further in post processing. There is a desirable effect of co-summing images into a deeper file, and that is an increase in the signal to noise ratio (S/N) of the final image as compared with the S/N of the individual summed images. This is the same advantage obtained by the stacking of images by Stacking software such as Autostakkert!, Registax or lxnstack for example. The S/N increases as the square root of the number of images stacked. If you stack 100 images, the resulting image will have a S/N 10 times greater than the individual images. Similarly, if you capture a hybrid long exposure by integrating, say 8 x 2s exposures, the resulting image will have the square root of 8 = 2.83 times the S/N of the individual 2s exposures.

It should be noted that summing identical images would neither increase the signal to noise ratio, nor give bit depth advantage. However the fact that every image is different due to random noise, seeing effects, thermal and other effects, means that some bit-depth advantage is achieved in addition to producing a brighter scaled image with more dynamic range and increased S/N ratio. It should also be noted that this is not the same as obtaining

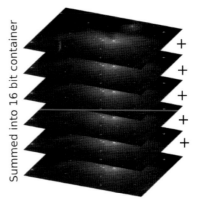

Plate 7: Synthesis of hybrid long exposures by summing a number of different 8 bit images into a 16 bit image file. This is stacking on a small scale during image capture. This technique, together with an appropriate scaling to the preview screen can also facilitate the preview of the faint object being imaged.

the data from a higher bit ADC in that the individual co-summed images are still individually digitised to an 8 bit range. Nevertheless, it does allow for an increase in the dynamic range and S/N ratio of the final image.

The camera used to capture the image in Plate 8 was an 8 bit DMK21AU. AS mounted on an 80mm ED refractor. The exposure of the 8 bit frames was set along with the gain and the gamma so that some nebulosity was visible and the central, trapezium region was not over-exposed. The final stacked image could be stretched so that the dimmer regions of nebulosity were revealed whilst not over brightening the central,

Plate 8: 23 hybrid long exposures of the Orion nebula, each of which was made by integrating 8 x 2s exposures into a 16 bit file using AstroDMx Capture for Linux. Total exposure time was 368s and the bit depth of each image was 11 bits with a range of 2,048 levels of brightness.

region of the nebula. This image has clearly benefited from off-chip integration and image stacking in that all parts of the nebulosity have been recorded while still resolving the trapezium within the nebulosity. Using a variety of scaling methods, the fainter parts of the image can be visualised during capture by AstroDMx Capture for Linux, whilst not affecting the collected data.

The DMK, DFK, DBK series of cameras are high speed 8 bit devices, intended primarily for solar system imaging, but are also capable of long exposures. There are available a variety of microscope cameras with a variety of video outputs such as Composite, HDMI, VGA and USB that can be obtained at the time of writing, from about £40. These are suitable for solar system astrovideography, for use with a video monitor or computer. One such camera is the 5MP Lucky Zoom USB camera which delivers a high quality video stream.

16 Bit Devices

Another class of camera intended for video astronomy uses 16 bit ADCs. These cameras give a greater advantage of dynamic range and long exposure capability. Examples are given by two Atic cameras. The first is the Atic Titan, a 16 bit planetary camera that is also able to do long exposures for deep sky work.

The Titan is able to capture 16 bit images at a respectable 15 frames per second. The advantage, of course, of such a 16 bit system, is that the subtleties in shades of brightness will be captured with more fidelity than with an 8 bit device, giving greater potential for processing the final image. The second example, the Atic Infinity, is a camera/software combination that allows the capture and stacking of long exposures, whilst viewing the capture process in real time. The software does a shift, de-rotate and stack on the fly. The stacked image is displayed on the screen and as the stack accumulates, the image noise can be seen to gradually disappear. Moreover, the shift and de-rotation means that an Alt-azimuth mount can be used as long as the exposures are short enough that there is no significant image rotation within an individual exposure.

In Plate 9, it appears as though the image is overexposed, with most of the nebula being saturated. However, this is just because the display has been set to display the fainter parts. The display can be set to various transformations of the displayed image without affecting the image data. It is the combination of camera and capture software that makes the Atic Infinity so powerful and so suitable for outreach work as well as serious deep sky imaging. The data can be captured as FITs files which can be variously, non-linearly stretched to reveal the whole nebula. The NASA, ESO, ESA software FITS Liberator is suitable software with which to do this and is free.

Plate 9: The Orion nebula being captured and displayed by the Atic Infinity camera and software.

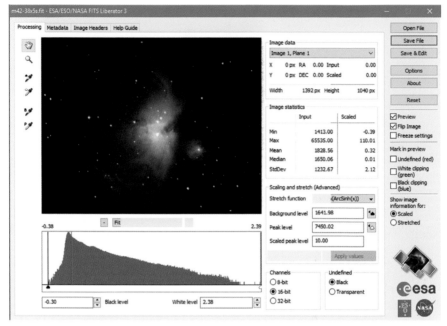

Plate 10: FITS Liberator stretching the data captured and shown in the Atic Infinity software in Plate 9. In this image, the stretch function and image scaling has been able to reveal all of the nebulosity whilst not over saturating the central, Trapezium region of the Orion Nebula.

12 Bit Devices

There is another class of device that has 12 bit ADCs. Some of the modern high speed planetary imagers fall into this category, such as the ZWO ASI120 MC/MM series. They produce video streams that are very fast depending on the region of interest being used. In these cameras, when a region of interest is selected, such as 320 x 240, suitable for containing a planet, only a small part of the sensor is read out and frame rates as high as 250 frames per second can be achieved. Our AstroDMx Capture for Linux software can move the region of interest across the chip in real-time to follow a planet that may be wandering slowly across the chip due to poor tracking. These cameras also have very sensitive CMOS sensors and can also be used to image some deep sky objects. They can produce streams of 12 bit RAW data in 16 bit format, where the 12 data bits are either placed in the top or the bottom 12 bits of a 16 bit container. They can also produce RAW 8 bit bayer video streams that can be debayered off camera or they can produce 8 bit RGB video streams.

Plate 11: M13, stack of
40 x 8s exposures with a
ZWOASI120 MC camera
at the Newtonian focus of
an f/5, 150 mm Newtonian
scope on an AZ, GOTO
mount.

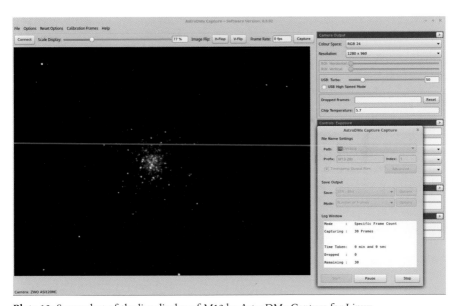

Plate 12: Screenshot of the live display of M13 by AstroDMx Capture for Linux.

Conclusion

In combination with capture software such as SharpCap for Windows, or AstroDMx Capture for Linux, cameras of various capabilities, costs and qualities can find a place in Video astronomy. With the appropriate camera, even deep sky objects can be viewed in real time.

If video files such as AVI, or the modern, superior format SER files are captured, then they can either be viewed at a future date as a record of the observing session, or they can be stacked by software such as Registax, Autostakkert!, lxnstack, Siril and others, to produce high quality still images that are capable of further enhancement by image manipulation software such as the Gimp, Photoshop and Paintshop Pro.

We will finish with a final word on the SER video file format. This is a superior movie format for a number of reasons. Firstly, the SER (SERial image container) player software can save out a SER file as another SER file or an AVI, containing a specified range of frames, or every nth frame. It can also save out the SER file as a sequence of images in various formats such as TIFFs. For the astronomer who likes to revisit an observing session by replaying the captured video, the SER player can enhance the video by changing brightness and contrast on the fly and the changes can be saved in a SER file to be viewed in the future. Or a selection of the frames can be saved as an animated GIF. The SER file is also a container that can hold 16 bit data, something that an AVI cannot do (although in principle it could). The SER file is therefore, a convenient single file container of many images that is, in my view, the file format of the future for video astronomy.

Miscellaneous

Some Interesting Variable Stars

Roger Pickard

I started observing variable stars many years ago, almost by accident.

I had built myself a 21.5 cm Newtonian telescope and had looked at many of the usual objects including the Moon, the planets, some open clusters and a few 'fuzzes'. But then my interest waned somewhat – until I joined my local astronomical society. They had sections – just like the British Astronomical Association (BAA) I was to learn in due course – and so I was to return to observing the moon and planets. But rather than just 'looking' at these objects, I was encouraged to report my observations to the BAA … an organisation that I was soon to join.

I was quite happy doing this, even after a friend invited me around to observe with his 30cm telescope when we also looked at a couple of variable stars. I had never looked at a variable before and I must say, although I thought it was interesting, it was not as good at looking at the Moon or planets! Not until I moved from suburbia to a much quieter area that was surrounded by trees, that is. There were not so many trees that I could not observe, and it was much darker than at my previous site. Naturally, I thought it should be great for observing, but it was not. Why? Trees give off some heat at night which makes the atmosphere 'tremble'. The upshot of this was that the 'seeing' is not so good, at least for observing the Moon and planets. However, I soon found that it did not really affect variable star observing, and at heart I was really an observer!

It was only a couple of years later that I got married and moved to another area that was in a valley. Again, the atmosphere was not that great for the Moon and planets, but it did not affect those variables, and so it came to pass that variable stars would become my main area of observing! I was greatly motivated by being a member of a local astronomical society and surrounded by so many like-minded people. I urge you to join your local society, as it can be a great motivator to do some 'proper' observing.

Different Types of Variable Star

So what is it about variable stars, apart from the fact that you can observe them at almost any time? For a start, they fit in with whenever you have time for observing. If you can only observe on, say, one night a week, but you can observe all night, then you choose those variables that we might refer to as 'one-nighters', these being stars that can go through most of their brightness variations in the course of a single night. The eclipsing binary Algol, or Beta (β) Persei, falls into this category (more about this star later).

Then we have stars that vary a bit more slowly, but which go through a cycle of variation over a week or two, examples being Beta (β) Lyrae and Delta (δ) Cephei. These would suit people who can observe for a short while on (nearly) every clear night.

Next we have slower variables whose brightness will change significantly over several months or a year. Examples include R Scuti and Z Ursae Majoris. These would suit people who can observe several times per month, but not necessarily on every clear night.

Stars whose brightness will change by many magnitudes over the course of a year would also suit people who can observe several times per month, but not necessarily on every clear night. They would especially appeal to observers who also possess a telescope. Examples here are the Mira-type variables, such as Mira itself (more about Mira later), R Hydrae and Chi (χ) Cygni.

Then there are stars that are totally unpredictable. These variables would mostly appeal to people who can observe on every clear night, although other observers may also enjoy monitoring them when their circumstances permit. R Coronae Borealis is a good example of this type, and novae would also fall into this category.

The trouble with variable stars is that they can often be difficult to find. Very few are as bright as any of the main planets and so they are much more likely to need some sort of optical aid to find them. Fortunately, the BAA's Variable Star Section (BAAVSS) has charts to help you find the more popular variables. Look for yourself by browsing the website at **www.britastro.org/vss**

Each chart shows the positions and magnitudes of a series of standard comparison stars (the 'sequence') against which you can make your magnitude estimates. The chart will also help you to locate the field of the variable star in relation to other stars. There are many star chart programmes available for the

PC nowadays which will help you to plan your own way of locating the variable before you go outside. Please remember to use the 'official' sequence to record your estimates.

We also have a mentoring scheme whereby you are put in touch with a competent observer to help you though your first stages of variable star observing, either personally or more probably via email.

Algol

Let us now take a look at a couple of these stars more closely, and here I recommend that you take a look at the BAAVSS Beginners Page. Towards the bottom, following the listing of suitable binocular and telescopic stars to observe at varying seasons, you will find the Eclipsing Binaries and Algol in particular. If you click on the star name in the table, it will bring up a comparison chart to help you make your observations (this also applies to other stars accessed via the 'Beginners' link). The comparison chart for Algol is included here along with predictions for timings of minimum brightness for the star throughout 2018.

But what type of star is Algol, and what do we mean by 'Eclipsing Binary'? To put it simply, an eclipsing binary is actually formed from two stars, one of which is fainter than its companion. The plane of their orbit is inclined to our line of sight and when the fainter star passes in front of, or eclipses, the brighter component, we see the overall magnitude of the pair fade as the light from the brighter component is temporarily obscured. In the case of Algol the reduction in brightness is well over a magnitude, so the dimming of the star should be easy to see. Algol does this every 2.87 days with the actual eclipse sequence taking just under 10 hours to fall and rise back to normal brightness. But you do not need to observe it for all that time just to see the changes in magnitude, although it can be useful to do that in order to see if your time of minimum matches that of the predictions! These can be seen in the table.

Beta (β) Persei (Algol): Magnitude 2.1 to 3.4 / Duration 9.6 hours

		h				h				h				h	
Jan	1	15.7		Feb	2	4.7	*	Mar	2	20.9		Apr	3	9.8	
	4	12.5			5	1.5	*		5	17.7			6	6.7	
	7	9.4			7	22.3	*		8	14.5			9	3.5	
	10	6.2			10	19.2			11	11.3			12	0.3	
	13	3.0	*		13	16.0			14	8.1			14	21.1	
	15	23.8	*		16	12.8			17	5.0			17	17.9	
	18	20.6	*		19	9.6			20	1.8	*		20	14.7	
	21	17.4			22	6.4			22	22.6	*		23	11.6	
	24	14.3			25	3.2	*		25	19.4			26	8.4	
	27	11.1			28	0.1	*		28	16.2			29	5.2	
	30	7.9							31	13.0					
May	1	2.0		Jun	1	15.0		Jul	3	4.0		Aug	3	16.9	
	3	22.8			4	11.8			6	0.8			6	13.7	
	6	19.6			7	8.6			8	21.6			9	10.6	
	9	16.5			10	5.4			11	18.4			12	7.4	
	12	13.3			13	2.2			14	15.2			15	4.2	
	15	10.1			15	23.1			17	12.0			18	1.0	*
	18	6.9			18	19.9			20	8.9			20	21.8	
	21	3.7			21	16.7			23	5.7			23	18.7	
	24	0.5			24	13.5			26	2.5			26	15.5	
	26	21.4			27	10.3			28	23.3			29	12.3	
	29	18.2			30	7.1			31	20.1					
Sep	1	9.1		Oct	2	22.1		Nov	3	11.0		Dec	2	3.2	*
	4	5.9			5	18.9			6	7.9			5	0.0	*
	7	2.7	*		8	15.7			9	4.7			7	20.8	*
	9	23.5	*		11	12.5			12	1.5	*		10	17.7	
	12	20.4			14	9.3			14	22.3	*		13	14.5	
	15	17.2			17	6.2			17	19.1			16	11.3	
	18	14.0			20	3.0	*		20	15.9			19	8.1	
	21	10.8			22	23.8	*		23	12.8			22	4.9	
	24	7.6			25	20.6			26	9.6			25	1.7	*
	27	4.4			28	17.4			29	6.4			27	22.6	*
	30	1.3	*		31	14.2							30	19.4	

Minima marked with an asterisk (*) are favourable from the British Isles, taking into account the altitude of the variable and the distance of the Sun below the horizon (based on longitude 0° and latitude 52°N)

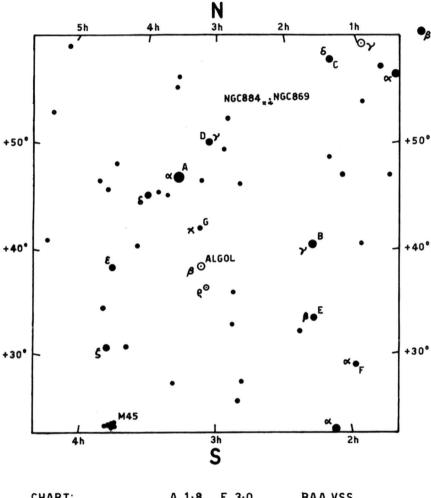

327·01   40° FIELD DIRECT

BETA PERSEI   03h 08m 10·1s +40° 57′ 20″ (2000)

CHART:	A 1·8	E 3·0	BAA VSS
NORTONS STAR ATLAS	B 2·1	F 3·4	EPOCH: 2000
SEQUENCE:	C 2·7	G 3·8	DRAWN: JT 19-06-11
HIPPARCOS VJ	D 2·9		APPROVED: RDP

Comparison chart for Algol

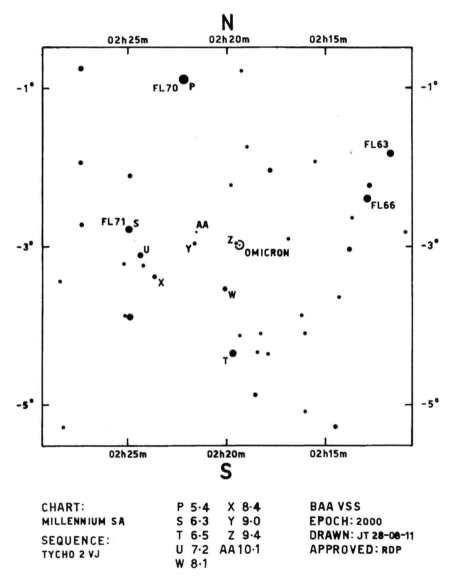

039·03   5° FIELD DIRECT

OMICRON CETI   02h 19m 20·8s   −02° 58′ 37″ (2000)

CHART:	P 5·4	X 8·4	BAA VSS
MILLENNIUM SA	S 6·3	Y 9·0	EPOCH: 2000
SEQUENCE:	T 6·5	Z 9·4	DRAWN: JT 28-08-11
TYCHO 2 VJ	U 7·2	AA 10·1	APPROVED: RDP
	W 8·1		

Mid-brightness comparison chart for Mira

Mira

Let us change variables now and take a look at Mira, or Omicron (o) Ceti, the prototype Long Period Variable (LPV). This is a completely different type of variable to Algol as it is a single star which we see varying due to pulsations taking place within the star itself. The period of Mira is around 332 days, so there will be no seeing this star rise and fall in one night! A typical maximum brightness for Mira is around magnitude 3 with a typical minimum of around 9.5 so, although you can see it without optical aid when it is at its brightest, you will need either a good pair of binoculars or a small telescope, and a dark observing site, to see it at minimum. The next maximum of Mira is likely to be around January 2018 with the following minimum occurring around August 2018. Charts for Mira are available to download from the Section's website. I say charts, as there is more than one to allow for observing the star at different magnitudes, although the mid-brightness chart is included here. It is when Mira gets brighter or fainter than the limit on this chart that you have to download one of the others.

If you do make observations of either of these stars (or any variable stars for that matter) please do report them to the BAAVSS. It is not essential that you become a member to submit observations, although I highly recommend that you do. You can also drop me an email (details on the website under 'Meet the VSS') at any time if you want to discuss anything to do with variable stars.

Some Interesting Double Stars

Brian Jones

The accompanying table describes the visual appearances of a selection of double stars. These may be optical doubles (which consist of two stars which happen to lie more or less in the same line of sight as seen from Earth and which therefore only appear to lie close to each other) or binary systems (which are made up of two stars which are gravitationally linked and which orbit their common centre of mass).

Other than the location on the celestial sphere and the magnitudes of the individual components, the list gives two other values for each of the double stars listed – the angular separation and position angle (PA). Further details of what these terms mean can be found in the article Double and Multiple Stars elsewhere in this volume.

Double-star observing can be a very rewarding process, and even a small telescope will show most, if not all, the best doubles in the sky. You can enjoy looking at double stars simply for their beauty, such as Albireo (β Cygni) or Almach (γ Andromedae), although there is a challenge to be had in splitting very difficult (close) double stars, such as the demanding Sirius (α Canis Majoris) or the individual pairs forming the Epsilon (ε) Lyrae 'Double-Double' quadruple star system.

The accompanying list is a compilation of some of the prettiest double (and multiple) stars scattered across both the Northern and Southern heavens. Once you have managed to track these down, many others are out there awaiting your attention …

Star	RA		Declination		Magnitudes	Separation	PA	Comments
	h	m	°	'		(arcsec)	°	
Beta[1,2] (β[1,2]) Tucanae	00	31.5	−62	58	4.36 / 4.53	27.1	169	Both stars again double, but difficult
Achird (η Cassiopeiae)	00	49.1	+57	49	3.44 / 7.51	13.4	324	Easy double
Mesarthim (γ Arietis)	01	53.5	+19	18	4.58 / 4.64	7.6	1	Easy pair of white stars
Almach (γ Andromedae)	02	03.9	+42	20	2.26 / 4.84	9.6	63	Yellow and blue-green components
32 Eridani	03	54.3	−02	57	4.8 / 6.1	6.9	348	Yellowish and bluish
Alnitak (ζ Orionis)	05	40.7	−01	57	2.0 / 4.3	2.3	167	Difficult, can be resolved in 10cm telescopes
Sirius (α Canis Majoris)	06	45.1	−16	43	−1.4 / 8.5			Binary, period 50 years, difficult
Castor (α Geminorum)	07	34.5	+31	53	1.93 / 2.97	7.0	55	Binary, 445 years, widening
Gamma (γ) Velorum	08	09.5	−47	20	1.83 / 4.27	41.2	220	Pretty pair in nice field of stars
Upsilon (υ) Carinae	09	47.1	−65	04	3.08 / 6.10	5.03	129	Nice object in small telescopes
Algieba (γ Leonis)	10	20.0	+19	50	2.28 / 3.51	4.6	126	Binary, 510 years, orange-red and yellow
Acrux (α Crucis)	12	26.4	−63	06	1.40 / 1.90	4.0	114	Glorious pair, third star visible in low power
Porrima (γ Virginis)	12	41.5	−01	27	3.56 / 3.65			Binary, 170 years, widening, visible in small telescopes
Cor Caroli (α Canum Venaticorum)	12	56.0	+38	19	2.90 / 5.60	19.6	229	Easy, yellow and bluish
Mizar (ζ Ursae Majoris)	13	24.0	+54	56	2.3 / 4.0	14.4	152	Easy, wide naked-eye pair with Alcor
Alpha (α) Centauri	14	39.6	−60	50	0.0 / 1.2			Binary, beautiful pair of stars
Izar (ε Boötis)	14	45.0	+27	04	2.4 / 5.1	2.9	344	Fine pair of yellow and blue stars
Omega[1,2] (ω[1,2]) Scorpii	16	06.0	−20	41	4.0 / 4.3	14.6	145	Optical pair, easy
Epsilon[1] (ε[1]) Lyrae	18	44.3	+39	40	4.7 / 6.2	2.6	346	The Double-Double, quadruple system with ε[2]
Epsilon[2] (ε[2]) Lyrae	18	44.3	+39	40	5.1 / 5.5	2.3	76	Both individual pairs just visible in 80mm telescopes
Theta[1,2] (θ[1,2]) Serpentis	18	56.2	+04	12	4.6 / 5.0	22.4	104	Easy pair, mag 6.7 yellow star 7 arc minutes from θ[2]

Star	RA		Declination		Magnitudes	Separation	PA	Comments
	h	m	°	'		(arcsec)	°	
Albireo (β Cygni)	19	30.7	+27	58	3.1 / 5.1	34.3	54	Glorious pair, yellow and blue-green
Giedi (α¹,² Capricorni)	20	18.0	−12	32	3.7 / 4.3	6.3	292	Optical pair, easy
Gamma (γ) Delphini	20	46.7	+16	07	5.14 / 4.27	9.2	265	Easy, orange and yellow-white
61 Cygni	21	06.9	+38	45	5.20 / 6.05	31.6	152	Binary, 678 years, both orange
Delta (δ) Tucanae	22	27.3	−64	58	4.49 / 8.7	7.0	281	Beautiful double, white and reddish

Some Interesting Nebulae, Star Clusters and Galaxies

Brian Jones

Object	RA		Declination		Remarks
	h	m	°	'	
47 Tucanae (in Tucana)	00	24.1	–72	05	Fine globular cluster, easy with naked eye
M31 (in Andromeda)	00	40.7	+41	05	Andromeda Galaxy, visible to unaided eye
Small Magellanic Cloud	00	52.6	–72	49	Satellite galaxy of the Milky Way
NGC 362 (in Tucana)	01	03.3	–70	51	Globular cluster, impressive sight in telescopes
M33 (in Triangulum)	01	31.8	+30	28	Triangulum Spiral Galaxy, quite faint
NGC 869 and NGC 884	02	20.0	+57	08	Sword Handle Double Cluster in Perseus
M34 (in Perseus)	02	42.1	+42	46	Open star cluster near Algol
M45 (in Taurus)	03	47.4	+24	07	Pleiades or Seven Sisters cluster, a fine object
Large Magellanic Cloud	05	23.5	–69	45	Satellite galaxy of the Milky Way
30 Doradus (in Dorado)	05	38.6	–69	06	Star-forming region in Large Magellanic Cloud
M1 (in Taurus)	05	32.3	+22	00	Crab Nebula, near Zeta Tauri
M38 (in Auriga)	05	28.6	+35	51	Open star cluster
M42 (in Orion)	05	33.4	–05	24	Orion Nebula
M36 (in Auriga)	05	36.2	+34	08	Open star cluster
M37 (in Auriga)	05	52.3	+32	33	Open star cluster
M35 (in Gemini)	06	06.5	+24	21	Open star cluster near Eta Geminorum
M41 (in Canis Major)	06	46.0	–20	46	Open star cluster to the south of Sirius
M44 (in Cancer)	08	38.0	+20	07	Praesepe, visible to naked eye
Carina Nebula (in Carina)	10	45.2	–59	52	NGC 3372, large area of bright and dark nebulosity
M104 (in Virgo)	12	40.0	–11	37	Sombrero Hat Galaxy to south of Porrima

Object	RA		Declination		Remarks
	h	m	°	'	
Coal Sack (in Crux)	12	50.0	−62	30	Prominent dark nebula, visible to naked eye
NGC 4755 (in Crux)	12	53.6	−60	22	Jewel Box open cluster, magnificent object
Omega (ω) Centauri	13	23.7	−47	03	Splendid globular in Centaurus, easy with naked eye
M51 (in Canes Venatici)	13	29.9	+47	12	Whirlpool Galaxy
M3 (in Canes Venatici)	13	40.6	+28	34	Bright globular cluster
M4 (in Scorpius)	16	21.5	−26	26	Globular cluster, close to Antares
M12 (in Ophiuchus)	16	47.2	−01	57	Globular cluster
M10 (in Ophiuchus)	16	57.1	−04	06	Globular cluster
M13 (in Hercules)	16	40.0	+36	31	Great Globular Cluster, just visible to naked eye
M92 (in Hercules)	17	16.1	+43	11	Globular cluster
M6 (in Scorpius)	17	36.8	−32	11	Open cluster
M7 (in Scorpius)	17	50.6	−34	48	Bright open cluster
M20 (in Sagittarius)	18	02.3	−23	02	Trifid Nebula
M8 (in Sagittarius)	18	03.6	−24	23	Lagoon Nebula, just visible to naked eye
M16 (in Serpens)	18	18.8	−13	49	Eagle Nebula and star cluster
M17 (in Sagittarius)	18	20.2	−16	11	Omega Nebula
M11 (in Scutum)	18	49.0	−06	19	Wild Duck open star cluster
M27 (in Lyra)	18	52.6	+32	59	Ring Nebula, brightest planetary
M27 (in Vulpecula)	19	58.1	+22	37	Dumbbell Nebula
M15 (in Pegasus)	21	28.3	+12	01	Bright globular cluster near Epsilon Pegasi
M39 (in Cygnus)	21	31.0	+48	17	Open cluster, good with low powers
M52 (in Cassiopeia)	23	24.2	+61	35	Open star cluster near 4 Cassiopeiae

M = Messier Catalogue Number
NGC = New General Catalogue Number

The positions in the sky of each of the objects contained in this list are given on the Star Charts printed earlier in this volume.

Astronomical Organizations

American Association of Variable Star Observers

49 Bay State Road, Cambridge, Massachusetts 02138, USA

www.aavso.org

The AAVSO is an international non-profit organization of variable star observers whose mission is to enable anyone, anywhere, to participate in scientific discovery through variable star astronomy. We accomplish our mission by carrying out the following activities:

- observation and analysis of variable stars
- collecting and archiving observations for worldwide access
- forging strong collaborations between amateur and professional astronomers
- promoting scientific research, education and public outreach using variable star data

American Astronomical Society

1667 K Street NW, Suite 800, Washington, DC 20006, USA

https://aas.org

Established in 1899, the American Astronomical Society (AAS) is the major organization of professional astronomers in North America. The mission of the AAS is to enhance and share humanity's scientific understanding of the universe, which it achieves through publishing, meeting organization, education and outreach, and training and professional development.

Astronomical Society of the Pacific

390 Ashton Avenue, San Francisco, CA 94112, USA

www.astrosociety.org

Formed in 1889, the Astronomical Society of the Pacific (ASP) is a non-profit membership organization which is international in scope. The mission of the ASP is to increase the understanding and appreciation of astronomy through the engagement of our many constituencies to advance science and science literacy. We invite you to explore our site to learn more about us; to check out our resources and education section for the researcher, the educator, and the backyard enthusiast; to get involved by becoming an ASP member; and to consider supporting our work for the benefit of a science literate world!

Astrospeakers.org
www.astrospeakers.org
A website designed to help astronomical societies and clubs locate astronomy and space lecturers which is also designed to help people find their local astronomical society. It is completely free to register and use and, with over 50 speakers listed, is an excellent place to find lecturers for your astronomical society meetings and events. Speakers and astronomical societies are encouraged to use the online registration to be added to the lists.

British Astronomical Association
Burlington House, Piccadilly, London, W1J 0DU, England
www.britastro.org
The British Astronomical Association is the UK's leading society for amateur astronomers catering for beginners to the most advanced observers who produce scientifically useful observations. Our Observing Sections provide encouragement and advice about observing. We hold meetings around the country and publish a bi-monthly Journal plus an annual Handbook. For more details, including how to join the BAA or to contact us, please visit our website.

Canadian Astronomical Society
Société Canadienne D'astronomie (CASCA)
100 Viaduct Avenue West, Victoria, British Columbia, V9E 1J3, Canada
www.casca.ca
CASCA is the national organization of professional astronomers in Canada. It seeks to promote and advance knowledge of the universe through research and education.   Founded in 1979, members include university professors, observatory scientists, postdoctoral fellows, graduate students, instrumentalists, and public outreach specialists.

Royal Astronomical Society of Canada

203-4920 Dundas St W, Etobicoke, Toronto, ON M9A 1B7, Canada

www.rasc.ca

Bringing together over 5,000 enthusiastic amateurs, educators and professionals RASC is a national, non-profit, charitable organization devoted to the advancement of astronomy and related sciences and is Canada's leading astronomy organization. Membership is open to everyone with an interest in astronomy. You may join through any one of our 29 RASC centres, located across Canada and all of which offer local programs. The majority of our events are free and open to the public.

Federation of Astronomical Societies

The Secretary, 147 Queen Street, SWINTON, Mexborough, S64 8NG

www.fedastro.org.uk

The Federation of Astronomical Societies (FAS) is an umbrella group for astronomical societies in the UK. It promotes cooperation, knowledge and information sharing and encourages best practice. The FAS aims to be a body of societies united in their attempts to help each other find the best ways of working for their common cause of creating a fully successful astronomical society. In this way it endeavours to be a true federation, rather than some remote central organization disseminating information only from its own limited experience. The FAS also provides a competitive Public Liability Insurance scheme for its members.

International Dark-Sky Association

darksky.org

The International Dark-Sky Association (IDA) is the recognized authority on light pollution and the leading organization combating light pollution worldwide. The IDA works to protect the night skies for present and future generations, our public outreach efforts providing solutions, quality education and programs that inform audiences across the United States of America and throughout the world. At the local level, our mission is furthered through the work of our U.S. and international chapters representing five continents.

The goals of the IDA are:

• Advocate for the protection of the night sky
• Educate the public and policymakers about night sky conservation
• Promote environmentally responsible outdoor lighting
• Empower the public with the tools and resources to help bring back the night

Royal Astronomical Society of New Zealand

PO Box 3181, Wellington, New Zealand
www.rasnz.org.nz
Founded in 1920, the object of The Royal Astronomical Society of New Zealand is the promotion and extension of knowledge of astronomy and related branches of science. It encourages interest in astronomy and is an association of observers and others for mutual help and advancement of science. Membership is open to all interested in astronomy. The RASNZ has about 180 individual members including both professional and amateur astronomers and many of the astronomical research and observing programmes carried out in New Zealand involve collaboration between the two. In addition the society has a number of groups or sections which cater for people who have interests in particular areas of astronomy.

Astronomical Society of Southern Africa

Astronomical Society of Southern Africa, c/o SAAO, PO Box 9, Observatory, 7935, South Africa
assa.saao.ac.za
Formed in 1922, The Astronomical Society of Southern Africa comprises both amateur and professional astronomers. Membership is open to all interested persons. Regional Centres host regular meetings and conduct public outreach events, whilst national Sections coordinate special interest groups and observing programmes. The Society administers two Scholarships, and hosts occasional Symposia where papers are presented. For more details, or to contact us, please visit our website.

Royal Astronomical Society

Burlington House, Piccadilly, London, W1J 0BQ, England

www.ras.org.uk

The Royal Astronomical Society, with around 4,000 members, is the leading UK body representing astronomy, space science and geophysics, with a membership including professional researchers, advanced amateur astronomers, historians of science, teachers, science writers, public engagement specialists and others.

Society for Popular Astronomy

Secretary: Guy Fennimore, 36 Fairway, Keyworth, Nottingham, NG12 5DU

www.popastro.com

The Society for Popular Astronomy is a national society that aims to present astronomy in a less technical manner. The bi-monthly society magazine Popular Astronomy is issued free to all members.

Our Contributors

Mike Frost is a Fellow of the Royal Astronomical Society and Director of the British Astronomical Association's historical section. His research interests centre on astronomers with connections to the English Midlands where he lives, and he lectures about them to astronomy societies around the country. Mike's day job is as a systems engineer for General Electric for whom he designs and commissions control systems for steel rolling mills around the world. Much of his article *Some Pioneering Lady Astronomers* was written in Belo Horizonte, Brazil.

Neil Haggath has a degree in astrophysics from Leeds University and has been a Fellow of the Royal Astronomical Society since 1993. A member of Cleveland and Darlington Astronomical Society for 35 years, he has served on its committee for 27 years. Neil is an avid eclipse chaser, clocking up five total eclipse expeditions so far to locations as far flung as Australia. Three of the expeditions were successful, and he plans to be in Jackson, Wyoming on 21 August 2017. In 2012, he may have set a somewhat unenviable record among British astronomers - for the greatest distance travelled (6,000 miles to Thailand) to NOT see the transit of Venus. He saw nothing on the day . . . and got very wet!

David A. Hardy, FBIS, FIAAA is the longest-established space artist, illustrating his first book, for Patrick Moore in 1954, and continuing to work with him on books and for 'The Sky at Night'. David is European Vice-President of the International Association of Astronomical Artists (IAAA) whose website is at **www.iaaa.org** and recipient in 2001 of the Rudaux Memorial Award for services to astronomical art. In March 2003 he became one of only a handful of artists to have had an asteroid named after him. David has also written or collaborated on many non-fiction books and has a published novel 'Aurora: A Child of Two Worlds'.

David Hardy's work, including originals, prints and DVDs, is available from AstroArt, 99 Southam Road, Hall Green, Birmingham B28 0AB (tel. 0121 777 1802), or go to his website **www.astroart.org**

Dr. David M. Harland gained his BSc in astronomy in 1977, lectured in computer science, worked in industry, and managed academic research. In 1995 he 'retired' in order to write on space themes.

David Harper, FRAS has had a varied career which includes teaching mathematics, astronomy and computing at Queen Mary University of London, astronomical software development at the Royal Greenwich Observatory, bioinformatics support at the Wellcome Trust Sanger Institute, and a research interest in the dynamics of planetary satellites, which began during his Ph.D. at Liverpool University in the 1980s and continues in an occasional collaboration with colleagues in China. He is married to fellow contributor Lynne Stockman.

Rod Hine was aged around ten when he was given a copy of 'The Boys Book of Space' by Patrick Moore. Already interested in anything to do with science and engineering he devoured the book from cover to cover. The launch of Sputnik I shortly afterwards clinched his interest in physics and space travel.

He took physics, chemistry and mathematics at A-level and then studied Natural Sciences at Churchill College, Cambridge. He later switched to Electrical Sciences and subsequently joined Marconi at Chelmsford working on satellite communications. This led to work in meteorological communications in Nairobi, Kenya and later a teaching post at the Kenya Polytechnic. There he met and married a Yorkshire lass and moved back to the UK in 1976. Since then he has had a variety of jobs in electronics and industrial controls, and recently has been lecturing part-time at the University of Bradford. Rod got fully back into astronomy in around 1992 when his wife bought him an astronomy book, at which time he joined Bradford Astronomical Society.

Brian Jones hails from Bradford in the West Riding of Yorkshire and was a founder member of the Bradford Astronomical Society. He developed a fascination with astronomy at the age of five when he first saw the stars through a pair of binoculars, although he spent the first part of his working life developing a career in mechanical engineering. However, his true passion lay in the stars and his interest in astronomy took him into the realms of writing sky guides for local newspapers, appearing on local radio and television, teaching astronomy and space in schools and, in 1985, leaving engineering to become a full time astronomy and space writer. His books have covered a range of astronomy and space-related topics for both children and adults and his journalistic work includes writing articles and book reviews for several astronomy magazines as well as for many general interest magazines, newspapers and periodicals. His passion for bringing an appreciation of the universe to his readers is reflected in his writing.

You can follow Brian on Twitter via **@StarsBrian** and check out the sky by visiting his blog at **www.starlight-nights.co.uk** from where you can also access his Facebook group Starlight Nights.

John McCue graduated in astronomy from the University of St Andrews and began teaching. He gained a Ph.D. from Teesside University studying the unusual rotation of Venus. In 1979 he and his colleague John Nichol founded the Cleveland and Darlington Astronomical Society, which then worked in partnership with the local authority to build the Wynyard Planetarium and Observatory in Stockton-on-Tees. John is currently double star advisor for the British Astronomical Association.

Neil Norman, FRAS first became fascinated with the night sky when he was five years of age and saw Patrick Moore on the television for the first time. It was the Sky at Night programme, broadcast in March 1986 and dedicated to the Giotto probe reaching Halley's Comet, which was to ignite his passion for these icy interlopers. As the years passed, he began writing astronomy articles for local news magazines before moving into internet radio where he initially guested on the Astronomyfm show 'Under British Skies', before becoming a co-host for a short time. In 2013 he created Comet Watch, a Facebook group dedicated to comets of the past, present and future. His involvement with Astronomyfm led to the creation of the monthly radio show 'Cometwatch', which is now in its second year. Neil lives in Suffolk with his partner and three children.

Damian Peach is a world renowned and multi award winning astrophotographer. Over his career his images have appeared in countless books, magazines and science papers, and he himself has also appeared numerous times on BBC television promoting astronomy and astrophotography as well as giving talks on the subject both in the UK and overseas.

Damian was awarded both the British Astronomical Association Merlin Medal and the Association of Lunar and Planetary Observers Walter Haas Award for outstanding contributions to planetary astronomy. In 2009 he was part of a record-setting team that produced the largest ground based image mosaic of the Moon ever taken. He featured in the acclaimed national Explorers of the Universe photographic exhibition at the Royal Albert Hall and also had his work featured at the Edinburgh Science Festival. His work has also been used by NASA and ESA to illustrate the importance and quality of amateur planetary images. In 2011 he was crowned overall winner of the Royal Greenwich Observatory Astrophotographer of the Year competition as well as being a prize winning finalist in 2012 and 2013. He also recently won 1st place in both the National Science Foundation's Comet ISON photo competition and in the solar system category of the APOTY 2016 competition.

Richard Pearson, FRAS, FRGS was born and raised in Nottingham and has worked on local newspapers as a journalist for over 20 years. A member of the British Astronomical Association and a Fellow of the Royal Astronomical Society, he has written several books on astronomy and is the presenter of the monthly internet TV program Astronomy & Space, now in its fourth year, and is a celebrity among members of astronomical societies worldwide.

Roger Pickard has been observing variable stars for more years than he cares to remember. Initially this was as a visual observer, but he then dabbled in photoelectric photometry (PEP) just before the advent of affordable CCDs. He now operates his telescope and CCD camera, which is located at the end of his garden, from within the warmth of his house. Roger has been the Director of the British Astronomical Association Variable Star Section since 1999 and can be contacted at **roger.pickard@sky.com** by anyone who needs help with variable star observing.

Peter Rea has had a keen interest in lunar and planetary exploration since the early 1960s and frequently lectures on the subject. He helped found the Cleethorpes and District Astronomical Society in 1969. In April of 1972 he was at the Kennedy Space Centre in Florida to see the launch of Apollo 16 to the moon and in October 1997 was at the southern end of Cape Canaveral to see the launch of Cassini to Saturn. He would still like to see a total solar eclipse as the expedition he was on to see the 1973 eclipse in Mali had vehicle trouble and the meteorologists decided he was not going to see the 1999 eclipse from Devon. He lives in Lincolnshire with his wife Anne and has a daughter who resides on the Gold Coast in Australia.

Lynne Stockman holds degrees in mathematics from Whitman College, the University of Washington and the University of London. She is a native of North Idaho but has lived in Britain for the past 25 years and is a visiting research student in astronomy at Queen Mary University of London. Lynne was an early pioneer of the world-wide web: with her husband David Harper, she created the web site **obliquity.com** in 1998 to share their interest in astronomy, family history and cats.

Steve Wainwright is a Fellow of the Royal Astronomical Society and a retired Senior Lecturer in Biological Sciences and Tutor in Astronomy from Swansea University. He is a Sky at Night veteran and founder of the QuickCam and Unconventional Imaging Astronomy Group (QCUIAG). Steve researches how serious astronomy can be done with modest equipment and has published a number of books and many scientific papers in learned journals. He is a past contributor to the Yearbook of Astronomy and has published a number of articles in various astronomy magazines and books. For several years he has collaborated in the development of astronomical software and is currently working with Nicola Mackin on Linux image capture software.

Glossary

Brian Jones

Altitude
The altitude of a star or other object is its angular distance above the horizon. For example, if a star is located at the zenith, or overhead point, its altitude is $90°$ and if it is on the horizon, its altitude is $0°$.

Apogee
This is the point in its orbit around the Earth at which an object is at its furthest from the Earth.

Asterism
An asterism is grouping or collection of stars located within a constellation that forms an apparent and distinctive pattern in its own right. Well known examples include the Plough (in Ursa Major); the False Cross (formed from stars in Carina and Vela); and the Summer Triangle, which is formed from the bright stars Vega (in Lyra), Deneb (in Cygnus) and Altair (in Aquila).

Asteroid
Another name for a minor planet.

Autumnal Equinox
The autumnal equinox is the point at which the apparent path of the Sun, moving from north to south, crosses the celestial equator.

Binary Star
See Double Star

Black Hole
A region of space around a very small and extremely massive collapsed star within which the gravitational field is so intense that not even light can escape.

Celestial Equator
The celestial equator is a projection of the Earth's equator onto the celestial sphere, equidistant from the celestial poles and dividing the celestial sphere into two hemispheres.

Celestial Poles
The north (and south) celestial poles are points on the celestial sphere directly above the north and south terrestrial poles around which the celestial sphere appears to rotate and through which extensions of the Earth's axis of rotation would pass.

The north celestial pole, the position of which is marked by the relatively bright star Polaris, lies in the constellation Ursa Minor (the Little Bear) and would be seen directly overhead when viewed from the North Pole. There is no particularly bright star marking the position of the south celestial pole, which lies in the tiny constellation Octans (the Octant) and which would be situated directly overhead when seen from the South Pole. The north celestial pole lies in the direction of north when viewed from elsewhere on the Earth's surface and the south celestial pole lies in the direction of south when viewed from other locations.

Celestial Sphere

The imaginary sphere surrounding the Earth on which the stars appear to lie.

Circumpolar Star

A circumpolar star is a star which never sets from any given latitude. When viewing the sky from either the North or South Pole, all stars will be circumpolar, although no stars are circumpolar when viewed from the equator.

Comet

A comet is an object comprised of a mixture of gas, dust and ice which travels around the Sun in an orbit that can often be very eccentric.

Conjunction

This is the position at which two objects are lined up with each other (or nearly so) as seen from Earth. Superior conjunction occurs when a planet is at the opposite side of the Sun as seen from Earth and inferior conjunction when a planet lies between the Sun and Earth.

Constellation

A constellation is an arbitrary grouping of stars forming a pattern or imaginary picture on the celestial sphere. Many of these have traditional names and date back to ancient Greece or even earlier and are associated with the folklore and mythology of the time. There are also some of what may be described as 'modern' constellations, devised comparatively recently by astronomers during the last few centuries. There are 88 official constellations which together cover the entire sky, each one of which refers to and delineates that particular region of the celestial sphere, the result being that every celestial object is described as being within one particular constellation or another.

Declination

This is the angular distance between a celestial object and the celestial equator. Declination is expressed in degrees, minutes and seconds either north (N) or south (S) of the celestial equator.

Deep Sky Object
Deep sky objects are objects (other than individual stars) which lie beyond the confines of our Solar System. They may be either galactic or extra-galactic and include such things as star clusters, nebulae and galaxies.

Double Stars
Double stars are two stars which appear to be close together in space. Although some double stars (known as *optical* doubles) are made up of two stars that only happen to lie in the same line of sight as seen from Earth and are nothing more than chance alignments, most are comprised of stars that are gravitationally linked and orbit each other, forming a genuine double-star system (also known as a *binary* star).

Eclipse
An eclipse is the obscuration of one celestial object by another, such as the Sun by the Moon during a solar eclipse or one component of an eclipsing binary star by the companion star.

Ecliptic
As the Earth orbits the Sun, its position against the background stars changes slightly from day to day, the overall effect of this being that the Sun appears to travel completely around the celestial sphere over the course of a year. The apparent path of the Sun is known as the ecliptic and is superimposed against the band of constellations we call the Zodiac (see Zodiac) through which the Sun appears to move.

Ellipse
The closed, oval-shaped form obtained by cutting through a cone at an angle to the main axis of the cone. The orbits of the planets around the Sun are all elliptical.

Ephemerides
Tables showing the predicted positions of celestial objects such as comets or planets.

Equinox
The equinoxes are the two points at which the ecliptic crosses the celestial equator (see also Autumnal Equinox and Vernal Equinox). The term is also used to denote the dates on which the Sun passes these points on the ecliptic.

Exoplanet
An exoplanet (or extrasolar planet) is a planet orbiting a star outside of our Solar System.

Galaxy
A galaxy is a vast collection of stars, gas and dust bound together by gravity and measuring many light years across. Galaxies occur in a wide variety of shapes and

sizes including spiral, elliptical and irregular and most are so far away that their light has taken many millions of years to reach us. Our Solar System is situated in the Milky Way Galaxy, a spiral galaxy containing several billion stars. Located within the Local Group of Galaxies (see below), the Milky Way Galaxy is often referred to simply as the Galaxy.

Inferior Planet
An inferior planet is a planet that travels around the Sun inside the orbit of the Earth.

International Astronomical Union (IAU)
Formed in 1919 and based at the Institut d'Astrophysique de Paris, this is the main coordinating body of world astronomy. Its main function is to promote, through international cooperation, all aspects of the science of astronomy. It is also the only authority responsible for the naming of celestial objects and the features on their surfaces.

Light Year
To express distances to the stars and other galaxies in miles would involve numbers so huge that they would be unwieldy. Astronomers therefore use the term 'light year' as a unit of distance. A light year is the distance that a beam of light, travelling at around 186,000 miles (300,000 km) per second, would travel in a year and is equivalent to just under 6 trillion miles (10 trillion km).

Local Group of Galaxies
This is a gravitationally-bound collection of galaxies which contains over 50 individual members, one of which is our own Milky Way Galaxy. Other members include the Large Magellanic Cloud, the Small Magellanic Cloud, the Andromeda Galaxy (M31), the Triangulum Spiral Galaxy (M33) and many others.

Galaxies are usually found in groups or clusters. Apart from our own Local Group, many other groups of galaxies are known, typically containing anywhere up to 50 individual members. Even larger than the groups are clusters of galaxies which can contain hundreds or even thousands of individual galaxies. Groups and clusters of galaxies are found throughout the universe.

Magnitude
The magnitude of a star is purely and simply a measurement of its brightness. In around 150BC the Greek astronomer Hipparchus divided the stars up into six classes of brightness, the most prominent stars being ranked as first class and the faintest as sixth. This system classifies the stars and other celestial objects according to how bright they actually appear to the observer. In 1856 the English astronomer Norman Robert Pogson refined the system devised by Hipparchus by classing a 1st magnitude star as being 100 times as bright as one of 6th magnitude, giving a

difference between successive magnitudes of $^5\ddot{O}100$ or 2.512. In other words, a star of magnitude 1.00 is 2.512 times as bright as one of magnitude 2.00, 6.31 (2.512 x 2.512) times as bright as a star of magnitude 3.00 and so on. The same basic system is used today, although modern telescopes enable us to determine values to within 0.01 of a magnitude or better. Negative values are used for the brightest objects including the Sun (-26.8), Venus (-4.4 at its brightest) and Sirius (-1.46). Generally speaking, the faintest objects that can be seen with the naked eye under good viewing conditions are around 6th magnitude, with binoculars allowing you to see stars and other objects down to around 9th magnitude.

Meridian
This is an imaginary line crossing the celestial sphere and which passes through both celestial poles and the zenith.

Messier Catalogue and References
References such as that for Messier 1 (M1) in Taurus, Messier 31 (M31) in Andromeda and Messier 57 (M57) in Lyra relate to a range of deep sky objects derived from the *Catalogue des Nébuleuses et des Amas d'Étoiles* (Catalogue of Nebulae and Star Clusters) drawn up by the French astronomer Charles Messier during the latter part of the eighteenth century.

Meteor
This is a streak of light in the sky seen as the result of the destruction through atmospheric friction of a meteoroid in the Earth's atmosphere.

Meteorite
A meteorite is a meteoroid which is sufficiently large to at least partially survive the fall through Earth's atmosphere.

Meteoroid
This is a term applied to particles of interplanetary meteoritic debris.

Milky Way
The faint pearly band of light crossing the sky which is the result of the combined light from the thousands of stars, the vast majority of which are too faint to be seen without optical aid, that lie along the main plane of our Galaxy as seen from Earth. Provided the sky is dark and clear the Milky Way is easily visible to the unaided eye, and any form of optical aid will show that it is indeed made up of many thousands of individual stars. Our Solar System lies within the main plane of the Milky Way Galaxy and is located inside one of its spiral arms. The Milky Way is actually our view of the Galaxy, looking along the main galactic plane. The glow we see is the combined light from many different stars and is visible as a continuous band of light stretching completely around the celestial sphere.

Nadir
This is the point on the celestial sphere directly opposite the zenith.

Nebula
Nebulae are huge interstellar clouds of gas and dust. Observed in other galaxies as well as our own, their collective name is from the Latin *'nebula'* meaning 'mist' or 'vapour', and there are three basic types:

- **Emission nebulae** contain young, hot stars that emit copious amounts of ultraviolet radiation which reacts with the gas in the nebula causing the nebula to shine at visible wavelengths and with a reddish colour characteristic of this type of nebula. In other words, emission nebulae *emit* their own light. A famous example is the Orion Nebula (M42) in the constellation Orion which is visible as a shimmering patch of light a little to the south of the three stars forming the Belt of Orion.
- The stars that exist in and around **reflection nebulae** are not hot enough to actually cause the nebula to give off its own light. Instead, the dust particles within them simply *reflect* the light from these stars. The stars in the Pleiades star cluster (M45) in Taurus are surrounded by reflection nebulosity. Photographs of the Pleiades cluster show the nebulosity as a blue haze, this being the characteristic colour of reflection nebulae.
- **Dark nebulae** are clouds of interstellar matter which contain no stars and whose dust particles simply blot out the light from objects beyond. They neither emit or reflect light and appear as dark patches against the brighter backdrop of stars or nebulosity, taking on the appearance of regions devoid of stars. A good example is the Coal Sack in the constellation Crux, a huge blot of matter obscuring the star clouds of the southern Milky Way.

Neutron Star
This is the remnant of a massive star which has exploded as a supernova.

New General Catalogue (NGC)
References such as that for NGC 869 and NGC 884 (in Perseus) and NGC 4755 (in Crux) are derived from their numbers in the New General Catalogue of Nebulae and Clusters of Stars (NGC) first published in 1888 by the Danish astronomer John Louis Emil Dreyer and which contains details of 7,840 star clusters, nebulae and galaxies.

Occultation
This is the temporary covering up of one celestial object, such as a star, by another, such as the Moon or a planet.

Opposition
Opposition is the point in the orbit of a superior planet when it is located directly opposite the Sun in the sky.

Orbit
This is the closed path of one object around another.

Penumbra
This is the area of partial shadow around the main cone of shadow cast by the Moon during a solar eclipse or the Earth during a lunar eclipse. The term penumbra is also applied to the lighter and less cool region of a sunspot.

Perigee
This is the point in its orbit around the Earth at which an object is at its closest to the Earth.

Planetary Nebula
Planetary nebulae consist of material ejected by a star during the latter stages of its evolution. The material thrown off forms a shell of gas surrounding the star whose newly-exposed surface is typically very hot. Planetary nebulae have nothing whatsoever to do with planets. They derive their name from the fact that, when seen through a telescope, some planetary nebulae take on the appearance of luminous discs, resembling a gaseous planet such as Uranus or Neptune. Probably the best known example is the famous Ring Nebula (M57) in Lyra.

Precession
The Earth's axis of rotation is an imaginary line which passes through the North and South Poles of the planet. Extended into space, this line defines the North and South Celestial Poles in the sky. The North Celestial Pole currently lies close to Polaris in Ursa Minor (the Little Bear), so the daily rotation of our planet on its axis makes the rest of the stars in the sky appear to travel around Polaris, their paths through the sky being centred on the Pole Star.

However, the position of the north celestial pole is slowly changing, this because of a gradual change in the Earth's axis of rotation. This motion is known as 'precession' and is identical to the behaviour of a spinning top whose axis slowly moves in a cone. Precession is caused by the combined gravitational influences of the Sun and Moon on our planet. Each resulting cycle of the Earth's axis takes around 25,800 years to complete, the net effect of precession being that, over this period, the north (and south) celestial poles trace out large circles around the northern (and southern) sky. This results in slow changes in the apparent locations of the celestial poles. Polaris will be closest to the North Celestial Pole in the year 2102, but it will then begin to move slowly away and eventually relinquish its position as the Pole Star. Vega will take on the role some 11,500 years from now.

Prime Meridian
The prime meridian is the meridian that passes through the vernal equinox.

Pulsar
This is a rapidly-spinning neutron star which gives off regular bursts of radiation.

Quasar
These are small, extremely remote and highly luminous objects which at the cores of active galaxies. They are comprised of a super-massive black hole surrounded by an accretion disk of gas which is falling into the black hole.

Right Ascension
The angular distance, measured eastwards, of a celestial object from the Vernal Equinox. Right ascension is expressed in hours, minutes and seconds.

Satellite
A satellite is a small object orbiting a larger one.

Shooting Star
The popular name for a meteor.

Sidereal Period
The time taken for an object to complete one orbit around another.

Solar System
The Solar System is the collective description given to the system dominated by the Sun and which embraces all objects that come within its gravitational influence. These include the planets and their satellites and ring systems, minor planets, comets, meteoroids and other interplanetary debris, all of which travel in orbits around our parent star.

Solstice
These are the points on the ecliptic at which the Sun is at its maximum angular distance (declination) from the celestial equator. The term is also used to denote the dates when the Sun passes these points on the ecliptic.

Spectroscope
An instrument used to split the light from a star into its different wavelengths or colours.

Spectroscopic Binary
This is a binary star whose components are so close to each other that they cannot be resolved visually and can only be studied through spectroscopy.

Spectroscopy
This is the study of the spectra of astronomical objects.

Star
A star is a self-luminous object shining through the release of energy produced by nuclear reactions at its core.

Star Colours

When we look up into the night sky the stars appear much the same. Some stars appear brighter than others but, with a few exceptions, they all look white. However, if the stars are looked at more closely, even through a pair of binoculars or a small telescope, some appear to be different colours. A prominent example is the bright orange-red Arcturus in the constellation of Boötes, which contrasts sharply with the nearby brilliant white Spica in Virgo. Our own Sun is yellow, as is Capella in Auriga. Procyon, the brightest star in Canis Minor, also has a yellowish tint. To the west of Canis Minor is the constellation of Orion the Hunter, which boasts two of the most conspicuous stars in the whole sky; the bright red Betelgeuse and Rigel, the brilliant blue-white star that marks the Hunter's foot.

The colour of a star is a good guide to its temperature, the hottest stars being blue and blue-white with surface temperatures of 20,000 degrees K or more. Classed as a yellow dwarf, the Sun is a fairly average star with a temperature of around 6,000 degrees K. Red stars are much cooler still, with surface temperatures of only a few thousand degrees K. Betelgeuse in Orion and Antares in Scorpius are both red giant stars that fall into this category.

Star Clusters

Although most of the stars that we see in the night sky are scattered randomly throughout the spiral arms of the Galaxy, many are found to be concentrated in relatively compact groups, referred to by astronomers as star clusters. There are two main types of star cluster – open and globular. Open clusters, also known as galactic clusters, are found within the main disc of the Galaxy and have no particularly well-defined shape. Usually made up of young hot stars, over a thousand open clusters are known, their diameters generally being no more than a few tens of light years. They are believed to have formed from vast interstellar gas and dust clouds within our Galaxy and indeed occupy the same regions of the Galaxy as the nebulae. A number of open clusters are visible to the naked eye including Praesepe (M44) in Cancer, the Hyades in Taurus and perhaps the most famous open cluster of all the Pleiades (M45), also in Taurus.

Globular clusters, as their name suggests, are huge spherical collections of stars. Located in the area of space surrounding the Galaxy, they can have diameters of anything up to several hundred light years and typically contain many thousands of old stars with little or none of the nebulosity seen in open clusters. When seen through a small telescope or binoculars, they take on the appearance of faint, misty balls of greyish light superimposed against the background sky. Although some form of optical aid is usually needed to see globular clusters, there are three famous examples which can be spotted with the naked eye. These are 47 Tucanae in Tucana (the Toucan), Omega Centauri in Centaurus (the Centaur) and the Great Hercules Cluster (M13) in the constellation Hercules.

Superior Planet
A superior planet is a planet that travels around the Sun outside the orbit of the Earth.

Supernova
Supernovae are huge stellar explosions involving the destruction of massive stars and resulting in sudden and tremendous brightening of the stars involved.

Synodic Period
The synodic period of a planet is the interval between successive oppositions or conjunctions of that planet.

Umbra
This is the main cone of shadow cast by the Moon during a solar eclipse or the Earth during a lunar eclipse. The term umbra is also applied to the darkest, coolest region of a sunspot.

Variable Stars
A variable star is a star whose brightness varies over a period of time. There are many different types of variable star, although the variations in brightness are basically due either to changes taking place within the star itself or the periodic obscuration, or eclipsing, of one member of a binary star by its companion.

Vernal Equinox
The vernal equinox is the point at which the apparent path of the Sun, moving from south to north, crosses the celestial equator.

Zenith
This is the point on the celestial sphere directly above the observer.

Zodiac
The band of constellations along which the Sun appears to travel over the course of a year. The Zodiac straddles the ecliptic and comprises the 12 constellations Aries, Taurus, Gemini, Cancer, Leo, Virgo, Libra, Scorpius, Sagittarius, Capricornus, Aquarius and Pisces.